WHERE DID THE
UNIVERSE
COME FROM?

AND OTHER
COSMIC QUESTIONS

WHERE DID THE
UNIVERSE
COME FROM?

AND OTHER
COSMIC QUESTIONS

OUR UNIVERSE, *from the* QUANTUM *to the* COSMOS

CHRIS FERRIE + GERAINT F. LEWIS

sourcebooks

Published by Sourcebooks
P.O. Box 4410, Naperville, Illinois 60567-4410
(630) 961-3900
sourcebooks.com

Library of Congress Cataloging-in-Publication Data

Names: Ferrie, Chris, author. | Lewis, Geraint F., author.
Title: Where did the universe come from? and other cosmic questions : our universe, from the quantum to the cosmos / Chris Ferrie, Geraint F. Lewis.
Description: Naperville, Illinois : Sourcebooks, 2021. | Includes bibliographical references and index.
Identifiers: LCCN 2021003815 (print) | LCCN 2021003816 (ebook) | (hardcover) | (epub)
Subjects: LCSH: Cosmology--Popular works.
Classification: LCC QB982 .F47 2021 (print) | LCC QB982 (ebook) | DDC 523.1--dc23
LC record available at https://lccn.loc.gov/2021003815
LC ebook record available at https://lccn.loc.gov/2021003816

Printed and bound in the United States of America.
LSC 10 9 8 7 6 5 4 3 2 1

To all who have wondered about the universe, from the very big to the very small and everything in between.

Contents

Part 3: The Quantum of Cosmos Future

Part 4: The Future of a Quantum Cosmos

Preface

The past several decades have blessed us with an amazing cast of scientists—too many to even name!—who have focused on communicating their research and discoveries to a wide audience of people outside the scientific community. Beyond possessing a gift to enrapture audiences of all ages and background, they tend to have one important thing in common: amazing photos of space. Surprisingly, if you dig a little bit into the expertise of many of these public figures, you'll find that not all of them are astronomers. Even scientists who may not work with telescopes draw us in with images of distant galaxies and illuminated clouds of interstellar dust.

Humans have always been stargazers, so it is no wonder we default to these inspiring photos of the heavens. The orderly beauty of the stars is where the word *cosmos* originates. Today, cosmos is synonymous with the universe. And while technically, this means *everything*, in practice, it means everything *out there*, beyond the confines of Earth. Cosmologists—the scientists who study the cosmos—typically have their eyes and telescopes trained on the stars.

The universe is unimaginably vast, so the word *cosmos* evokes an enormous sense of scale. The cosmos, while technically including things like ants, grains of sand, and individual atoms, is studied in such a way as to ignore the small details. When viewed through this lens of vastness, which focuses on things like planets, stars, and black holes, the science of the cosmos seems to stick to a well-ordered set of physical theories and laws. From this point of view, the universe is a well-oiled machine, predictable and constant. We understand basically how it works.

One of the biggest surprises in the history of science, however, is that the seeming order of the cosmos does not apply at every scale. We might have expected that something big but far away is the same as something small but very close. A telescope magnifies distant objects. A microscope magnifies tiny objects. But when we look deep into the world of the very small, we find an unfamiliar world seething with the unpredictable and counterintuitive, in many ways very different from the functions of the cosmos. In the very small, we see the *quantum* world.

Quantum is a word that describes the branch of modern physics developed at the turn of the twentieth century to explain experiments that investigated things much smaller than even our best microscopes can see. This new world was that of atoms. And although this *quantum world* is impossible for us to directly observe, it provided the basis for our understanding of the elements, chemistry, and eventually the stars themselves.

Not only is quantum physics the basis of all modern technology, it is the thread that connects all scientific disciplines. The theory itself, however, is notoriously difficult to comprehend. Our task here is not to provide the reader with a working knowledge of quantum physics—so as to build their own technology or theories— but to provide an appreciation for the science that connects us to the cosmos.

This book is about the quantum and the cosmos, the two extremes of human understanding. The quantum is the world of the very small, of atoms and electrons, with fundamental forces and fundamental particles, the building blocks of everything. The cosmos is the whole shebang, a universe of trillions of stars and galaxies, expanding space from a fiery birth to an unending future. The two might sound quite distinct, but through this book, we will show that they are intricately intertwined, with the life of the universe on the largest scales tied up with the action of the quantum on the smallest.

We will explore our current understanding of the universe and see how it is the very small that informs our theories of the very large. We will venture into the distant past and speculate on the far future—a cosmic show viewed through the lens of the quantum world. And although the quantum and the cosmos could not be further separated in scale, it is when they are brought together that the true beauty of the heavens is revealed.

Chris Ferrie and *Geraint F. Lewis*
Sydney, Australia, 2021

The Quantum and the Cosmos

They say that the most incredible thing about the universe is that we can understand it. But of course, we don't completely understand it—not yet anyway.

There are many things about the universe that remain dark and mysterious. But for a slightly evolved ape whose civilization is measured in thousands of years rather than the billions of years that have marked the passage of cosmic time, humans have done quite well!

Over the last few centuries, we've successfully unraveled much of the language of the universe. We've discovered that the rules that govern how things change and interact are not written in words but in mathematical equations. From the first steps of Galileo, Kepler, and Newton four hundred years ago, the universe has steadily given up its mathematical secrets. Seemingly mysterious phenomena—such as electricity and magnetism, matter and light, heat and energy—were explored, defined, explained, and finally articulated in the beauty of mathematical equations.

By the end of the nineteenth century, it looked like the end might be in sight. The great Lord Kelvin is rumored to have said that there was "nothing new to be learned in science," and all that was left to do was to make measurements at higher and higher precision.[1]

But this cozy view of the scientific universe was about to come tumbling down. The start of what became a series of revolutions in science can be traced to the turn of the twentieth century, when a forty-two-year-old German physicist was trying to make sense of the world.

Max Planck was trying to understand why things glowed when you heated them. Of course, many things simply burn—a chemical reaction that turns one substance into another. But if you have ever seen a blacksmith shoe a horse or seen a poker in a hot fire, you know that heated metal glows. At first, it glows with a rosy red, but as it gets hotter, metal can become white hot. What is the source of the color of heated metal?

Planck was not trying to explain the colors of heated metal in some general wishy-washy terms. No, he wanted to describe the observed color of hot metal precisely—why so much red compared to blue? Remember, when you heat up something, it turns red and then white hot. The question your inner child is yearning to answer is *why?*

Planck was not the first to try answering this question, but everyone who came before him had failed. They derived their mathematical relationship for the color of hot metal using the laws of the universe as they understood them. They knew the light was emitted when tiny electrical charges—which we now know are electrons—inside the metal jiggled around, oscillating back and forth. These jiggling charges emit light. Heating the metal gave these little charges more energy, so they jiggled more furiously, emitting more light. Scientists realized that the emitted color is implicitly tied to jiggling charges, so determining how the energies from the heat made the charges oscillate was key to their calculations.

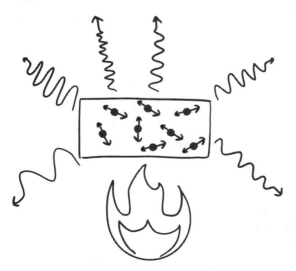

Unfortunately, their mathematics failed. Scientists could correctly account for the amount of red light—light with lower energy and longer wavelengths. Blue light, however, has more energy than red light, and their mathematics predicted there should be more blue light than red light. But they also predicted there should be even higher energy radiations, such as ultraviolet, X-rays, and gamma rays, than blue light, and this was simply not seen in experiments. This "ultraviolet catastrophe" marked a failure of our understanding of the physical world.

Planck, too, was on the brink of failure when he tried something radical. This was surprising because Planck—as described in his obituary by fellow physicist Max Born—was a conservative man, skeptical of speculation. Being radical was not in his nature, but he felt that he had no option.[2] He concluded that the laws of physics, as he understood them, could not solve the problem of the color of heated metal.

The Quantum Hypothesis

Planck's revelation was to consider the jiggling of the charges as being *discrete*—coming in indivisible chunks. Discrete might seem like a strange word to use, but it's easy to think about in terms of money. Imagine that you have a stack of one-dollar bills. If I ask you count out an amount of money using this stack, you are always going to get a whole number of dollars: $0, $1, $2... With just a stack of one-dollar bills, you are never going to count out $1.23, unless you start tearing up the bills, which is a bad idea!

Planck assumed that the jiggles of charges in a hot body were discrete, like counting with the stack of one-dollar bills, and jiggles in between these discrete amounts were forbidden. A word like *forbidden* might sound a little weird when we're talking about physical laws and theories, but what we mean is that Planck wrote these rules into his mathematics to see what the consequences would be. He did not know *why* the rules would be that way.

To his astonishment, the new rules worked! The color of hot metal was accurately described by the mathematics of Planck's particular oscillating charges. The problem was that this new approach flew in the face of established physics. Over 250 years prior, Newton had brought calculus to bear on the physical world. The incredible success of calculus ingrained in the minds of all scientists that the world and everything in it were *continuous*—everything could be split in half, and those halves could be split in half, and again and again and again, forever. That there would be some stopping point where everything eventually became discrete—as Planck was suggesting—was unacceptably arbitrary and inelegant in a world that seemed to mirror the perfect mathematical beauty of the infinite.

Planck was bemused by this finding and wondered if he had stumbled across a mathematical sleight of hand. He felt that maybe, if he dug deeper, he would find that this trick was actually built upon the established laws of physics, and everything would be congruent with the universe as science understood it. But eventually, it became obvious to Planck and other physicists that this was not the case. Physically, changes of energy on the very small scale come in little chunks, or *quanta*, rather than being continuous. Through his work on the glow of metal, Planck had, unbeknownst to himself, taken the first steps into what we now know as *quantum theory*.

Over the following decades, physicists elaborated on this idea of the quantum, and at every turn, the small scale defied the universe defined by the exactness of the Newtonian forces and motion of everyday life. The quantum world, generations of scientists discovered, was governed by esoteric mathematics and chance. Perhaps the abstractness of these concepts contributed to their delayed appreciation. However, once the experimental discoveries built upon the new physics started rolling in, the upheaval of the scientific community was swift. Without quantum physics, we could *maybe* power the world with coal-fired electricity. But with quantum physics, we now have the terrifying ability to destroy it. Quantum physics gives us the description of nature that we have built our modern technological world upon.

It was always assumed that the world—be it quantum or otherwise—played out on the background stage of space and marched in tune with the ticking of universal time. But as it turned out, these ideas were due for a revolution as well.

Shedding a Light on Space and Time

There was another scientist working at the birth of quantum mechanics who provided radical insights into the nature of light and matter. He eventually came to object to the emerging consensus on the abstract nature of the quantum world. Though he played a major role in developing the field of quantum physics, he spent his later years arguing with its leading champions. But that's not why he is in this story. He is here because he turned his eyes skyward and revolutionized our understanding of the heavens. The scientist's name was Albert Einstein.

Einstein, too, wondered about the fundamental nature of the universe. He thought not about the atoms and light that inhabited it but rather about the space and time in which they played. In the views of his predecessors, space and time were rigid and immutable—a stage for physics to play out against according to the universal laws of motion. Einstein's ideas changed all this, making use of the Gedankenexperiment—or thought experiment—technique, which he is now famous for. Leading up to 1905, which is known as Einstein's "miracle year," his thought experiments focused on the movement of light and how different observers might perceive it.[3]

Way back in the sixteenth century, Galileo had demonstrated that motion is relative. There was no experiment you could do—such as throwing a ball or observing a bee fly—that could reveal whether you are sitting in your chair at home or on a ship sailing smoothly across a glassy sea. But another person, observing from a different state of motion *relative* to yours, would surely notice the difference. However, that person has no privileged position, because they would not be able to tell who was moving, them or you! If you have ever been sitting in a car and felt the sensation that you were moving backward because the car next to you started moving forward, you have *felt* the relativity of motion. For Galileo, *all* motion was relative, and there were no absolutes. But Galileo did not know about the nature of light and certainly would not have guessed that it would change our ideas of motion.

What is light? This question was answered by the Scottish physicist James Clerk Maxwell in the mid-1800s. His starting points were two seemingly distinct concepts: electricity and magnetism. Maxwell showed that they are implicitly related and could be united in four interrelated equations, compact enough to be displayed on the T-shirts of science fans everywhere. Maxwell wrote down his famous equations, and God said "Let there be light" is the type of nerdy humor this has spawned. The in-joke relies on knowledge of the fact that Maxwell's equations are also the laws of light.

MAXWELL'S EQUATIONS

$$\nabla \cdot \vec{D} = \rho$$

$$\nabla \cdot \vec{B} = 0$$

$$\nabla \cdot \vec{E} = -\frac{\partial \vec{B}}{\partial t}$$

$$\nabla \cdot \vec{H} = \vec{J} + \frac{\partial \vec{D}}{\partial t}$$

Maxwell's equations are very powerful, functioning as one small set of mathematical equations to explain the complete nature of electricity and magnetism. But he realized there was something deeper buried in his mathematics as well. Maxwell's equations described space as being filled with fields, one electric and one magnetic. It is through these fields that electric charges and electric currents *communicate*, attracting and repelling via the force of electromagnetism.

Maxwell realized that a changing magnetic field will produce an electric field, and a changing electric field will produce a magnetic field, and so on. There was nothing in the equations that required these periodic changes to stop, and they could in principle propagate themselves as a wave moving through empty space. He decided to check how fast these *electromagnetic waves* would move, and to his astonishment, their speed was 299,792,458 meters per second (m/s), exactly matching the speed of light. Maxwell deduced that light itself was an electromagnetic wave.

As well as the optical light we can see, Maxwell predicted that there must be other electromagnetic waves that are invisible to us. Electromagnetic waves are defined by the length of the waves, and our eyes are sensitive to waves as small as about 0.4 of a thousandth of a millimeter, which we see as blue light. The longest waves our eyes can sense are about twice as long, which we see as red light. But beyond this narrow window, Maxwell reasoned, there must be waves of longer and shorter lengths that our eyes cannot see. In the late 1800s, with the detection of radio waves by Heinrich Hertz and X-rays by Wilhelm Roentgen, Maxwell's full hypothesized spectrum of electromagnetic waves was confirmed.[4]

Maxwell's equations of electromagnetism were a great success, but Einstein was looking for something more. He knew that the mathematical description of electromagnetic waves revealed their speed through a vacuum as a blistering 300,000 km/s! What bothered him was that nothing told him what this speed was relative to. Other physicists had suggested that space was full of a "substance" in which electromagnetic waves rolled, like waves on the ocean. They called this invisible electromagnetic sea the *aether*. However, experiments built to confirm the presence of the aether continually came up short, suggesting electromagnetic waves travel through empty space.

Einstein's brilliance was to speculate that the speed of light was relative to everyone and everything and that everyone would measure it to be the same 300,000 km/s. It would be the one absolute in an otherwise relative world. But this was impossible in Newton's universe, where all speeds were relative and everyone should measure a different speed of light. Certainly, someone traveling just shy of 300,000 km/s, alongside a beam of light, would watch the light inch away from them, right? Not according to Einstein. That person would still measure light speeding away at 300,000 km/s relative to them. Clearly, to make this all work out, something profound would have to give.[5]

And what gave was the notion of rigid and immutable space and time. These concepts would have to be abandoned and replaced with something more malleable. With this change, everyone could measure light as moving at the same speed. The consequences were that everyone's clock would have to tick at a different rate, and everyone's ruler would be a different length. No longer could observers agree on the distance between two points or how long something took to happen! With the announcement of his special theory of relativity, Einstein seemed to be throwing out the entire background of the physical universe—and he wasn't finished there.

The Gravity of the Situation

Einstein realized that the dominant force in the universe, gravity, just didn't fit in to his special relativistic picture. Isaac Newton had defined the mathematical form of gravity in the seventeenth century, and it had worked extremely well until that point. But Newton's formula—his so-called *universal law of gravity*— depended upon the distance between masses, and if no two distance measurements agree, just which one do you use? It took Einstein a decade's worth of effort to develop his solution, the *general theory of relativity*.

Einstein thought about someone falling under the force of gravity. Imagine that person sitting in a room surrounded by everyday objects, such as a table, chairs, plates, cups, and saucers. If the entire room is falling under the force of gravity, the person and all these objects would appear to hang weightlessly in the air within the room. Einstein realized that from the falling person's point of view, gravity would have effectively disappeared!

Einstein's thought experiment spurred him on to incorporate the phenomenon of gravity into his malleable vision of space and time. The mathematics were fiendish and the going difficult. Finally, in 1915, he succeeded. To incorporate gravity into the theory of relativity, Einstein showed that space and time had to be truly bendy. The ticking of a clock and the length of a ruler depend on where the objects are relative to objects with mass, the source of gravity.

The consequences of gravity being related to curved space and time were revolutionary. For a long time, astronomers had noticed that the orbit of Mercury, the closest planet to the Sun, would sometimes wander away from the predictions of its path according to Newtonian gravity. Einstein's new mathematics could account for this wandering. His theory also predicted that the path of light through the universe was not a simple straight line but would be deflected in the presence of a massive object. It was the detection of this *gravitational lensing* during a solar eclipse in 1919 that propelled Einstein to international acclaim.

We experience the influence of Einstein's theory of relativity every day without really having to think about it. For example, it is a vital aspect of the Global Positioning System (GPS), which relies on a synchronized network of clocks. The GPS needs to pass messages across large distances and must therefore know the *when* as well as the *where*. Without accounting for the relative bending of space and time between clocks on satellites and clocks on Earth, the time shown on those clocks would rapidly drift into uselessness, and we'd all end up in locations we didn't expect!

But for the purposes of our story, the greatest success of the theory of relativity was its description of the entire history of the universe as a whole. Einstein was one of the first to try and derive the mathematics of the cosmos, of all space and all time. In his mind and set against the backdrop of the 1800s, the universe was static and unchanging, and his mathematical models of the universe reflected this assumption. It was a little-known Russian mathematician, Alexander Friedmann, who, in 1922, published the idea that the universe was actually a dynamic and evolving place. From that point on, modern cosmology—the science of the origin and development of the universe—evolved rapidly. Edwin Hubble discovered that all other galaxies seemed to be moving *away* from our own—the universe was not only changing, it was *expanding*. At the same time, Georges Lemaître reasoned that if the universe is expanding, it must have started at some time in the finite past and that there must have been a cosmic birth. He dubbed this event the *primeval atom*, but it would be soon be better known as the *Big Bang*.

The Two Pillars

By the mid-twentieth century, significant progress had been made toward understanding the nature of the universe, and the blueprint for modern physics had been laid out. The problem was that it was actually two distinct blueprints, quite different from each other. Gravity was referenced in the language of Einstein's general theory of relativity in terms of curvy and bendy space, whereas the other forces of nature—electromagnetism and the subatomic forces—were encoded in the discreteness of quantum mechanics.[6]

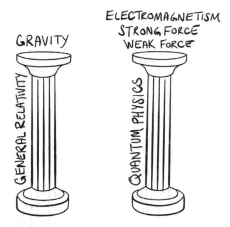

If you pick up a basic physics textbook, you will see this distinction for yourself. Chapters on quantum mechanics are often quite distinct from the chapters on relativity and gravity. Different characters appear, with Bohr, Pauli, and Schrödinger central to quantum mechanics, while Newton, Einstein, and Schwarzschild appear in the story of gravity.

Wandering around any physics department will reveal the same distinction. You might find a corridor with posters for conferences on quantum computing, advanced materials, or superconductors—all areas defined by the rules of quantum physics. Other corridors might be adorned with posters on cosmology, dark matter and dark energy, or even the early universe. Here, gravity dominates, and the language of relativity is spoken.

While these physicists may chat about football and mortgages over departmental coffee, in terms of science, they appear to be speaking completely different languages!

Being built on two distinct pillars—relativity and quantum mechanics—there appears to be a schism in modern physics. The mathematics of relativity is used to describe the physics of the large and massive—planets, stars, and galaxies—while quantum mechanics reigns over the very small, the world of electrons and particles. These domains appear so distinct that if you focus on one thing, you can often neglect the other altogether. The astronomer who studies the motion of planets and comets can rely on the equations of gravity and ignore everything else. The physicist trying to build a quantum computer out of a series of individual atoms, however, can happily ignore the puny gravitational pulls between them.

The existence of these two separate pillars is a concern for physicists and has been a driving force in their search for a single "theory of everything" that completely underwrites the universe. We will return to this point in the closing chapter of this book and look at the challenges and solutions in modern fundamental physics.

The separation of modern physics into the separate worlds of the quantum and gravity definitely makes understanding the universe a challenge. But this does not mean that modern physics has failed. Where we can make these two ideas work together, the cosmos has yielded its innermost secrets, from its fiery beginnings to the cold, unending future that lies ahead. Exploring these is the goal of this book.

We're going to take a journey through the life of the cosmos, wondering about its birth and the forces that shaped its very being. We'll uncover the lives of stars and the formation of the elements. And we will ponder what awaits the universe in the long, dark future ahead. Through all this, we will find gravity playing its dominant role, defining the expansion of the universe and squeezing matter into stars. But in understanding the universe, gravity is not enough, and the role that other forces play cannot be ignored. In fact, we will find quantum mechanics at every turn and playing the defining role.

We will see that in order to really understand our place in the universe, we cannot separate the quantum and the cosmos.

PART 1

The
QUANTUM
of COSMOS
PAST

Where did the universe come from?

When the night is dark, the sky is lit with thousands of stars. As we gaze upon its glory, it is easy to imagine that the universe has always been this way. But we know this is an illusion. In the life span of the universe, human lives and civilizations pass in the blink of an eye. If we were around for long enough, over millions and billions of years rather than the mere thousands that have passed since humans planted the first crops and built the first cities, we would see that we live in an evolving and changing universe.

Cosmology is the study of the evolution of the universe. While people have looked into the skies for meaning from the earliest beginnings of humanity, cosmology has only become a true science over the last century. Advances in telescopes have opened up the heavens, revealing a universe much larger and richer than we could have ever imagined. Our Sun is one star of hundreds of billions in the Milky Way galaxy, whose light shines across the sky from horizon to horizon. And the Milky Way is just one of possibly trillions of galaxies visible to our most powerful telescopes.[1]

As the universe came into sharper view through our telescopes, another revolution was underway. In the early part of the twentieth century, Einstein put the finishing touches to his general theory of relativity, pushing aside Newton's mathematics of gravity, which had reigned for three hundred years. This new view of the universe, where gravity is encoded in the warping and bending of space and time, is starkly different from the rigid space and time of Newton but completely subsumed the predictive power of his picture of gravity and gave so much more. Within the mathematics of relativity lay explanations of supercondensed stars, black holes, wormholes, and the rippling and waving of space and time themselves.

Also buried there was the mathematics of the universe, and what a universe it was! Not a static and unchanging cosmos, as Einstein had initially imagined, but a dynamic universe that was constantly evolving. As the famous astronomer Edwin Hubble peered through his telescope in the 1920s, he observed this new understanding, seeing galaxies rush away from one another as the universe expands.[2]

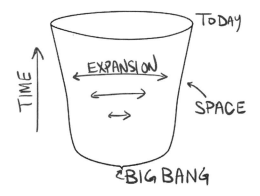

It didn't take long for people to realize the amazing implication of an expanding universe. If galaxies will be farther apart tomorrow, they were closer together yesterday. Looking back further and further into the past, galaxies must have been packed closer and closer together. At a point about fourteen billion years ago, the distances between all the galaxies become nothing, a starting point for the expansion we see today. This means there was a universal birth, the existence of a day without a yesterday.

With things squeezed together, the universe of the past must have been hotter and denser than the universe today. In its initial moments, it must have been extremely dense and extremely hot. This fiery birth was named the *Big Bang* by eminent astronomer Fred Hoyle, a disparaging term as he could not accept a universe with a beginning. Hoyle had his own ideas of a universe that expands but has lasted forever, something known as the *steady-state theory*.

Despite its inauspicious naming, the idea of the Big Bang stuck, and the idea of an expanding universe born at a finite time in the past has become the best explanation we have for the cosmos we observe around us.

While Einstein's general relativity led us to the Big Bang, we need more physics to describe the complex interactions underway when the universe was immensely hot and dense. As well as the intense pull of gravity, the vigorous collisions between the basic building blocks of matter, elementary particles like *electrons* and *quarks*, mean we cannot neglect the other fundamental forces: electromagnetism, the strong force, and the weak force. We will revisit these forces numerous times in this book, but for now, all we need to know is that each of these three phenomena is described by the laws and language of quantum mechanics.

In the earliest stages of the universe, gravity and the other forces were vying for dominance. In describing the universe, neither quantum mechanics nor general relativity can be ignored. Both must be used on an equal footing. But we don't quite know how to reconcile these two distinct frameworks in a way that brings them together seamlessly.

To explore the earliest stages of the universe, we have to stick the various bits of mathematics together in a rather haphazard way in an attempt to join the four fundamental forces (gravity, electromagnetism, strong force, and weak force) into something we think—we hope—works.

There is more than one way of sticking the pieces of mathematics together, and we don't really know if any approach is a good approximation of reality in the extreme conditions of the earliest epochs of the universe. Eventually, we reach a point, peering back earlier into the history of the universe, where we know this Frankenstein mathematics simply cannot work. We end up stuck, facing a wall in our physical theories and unable to explore any further. This prevents us from revealing the mechanics of the birth of our universe and answering the fundamental question *where did the universe come from?*

However, we can still ponder and imagine what an answer might look like. To do this, we're going to have to think a little bit about nothing. Nothing at all! What could be easier?

Thinking about Nothing

"Nothing" is a topic that both physicists and philosophers argue about. One kind of nothing, a chunk of space devoid of any matter or radiation, is a simple kind of nothing. But there is another kind of nothing, where you also strip away the space and time themselves. This second nothing is harder to imagine, so let's start thinking about just an empty piece of existing space and time.

Imagine stepping into the boots of a spacewalking astronaut

and gazing into the universe. You might catch a glimpse of the void of space. Peering into the nothingness might present us with an unmatched sense of existential angst, but solace comes from the most unlikely of sources: quantum physics. For even empty space itself seethes with particles popping in and out of existence. These are called *quantum fluctuations*.[3]

Particles popping in and out of existence sounds like yet another weird concept dreamed up by scientists to confuse everyone else. But the underlying structure of quantum mechanics demands their presence, and while we cannot see them directly, we can measure their influence on the world around us.

Quantum fluctuations—as the name suggests—are fleeting and fickle, but they have always been and thus will forever be. In an eternal universe, the only constant is the never-ceasing movement of quantum energy.[4] Yet the very seeds of our understanding of these quantum fluctuations were sown less than one hundred years ago.

The first glimpse of quantum mechanics—and quantum fluctuations—occurred on the treeless island of Heligoland in the North Sea. There, escaping hay fever in his native Germany in 1925, theoretical physicist Werner Heisenberg conceived the basic mathematics of the newly developing quantum theory. Until then, physicists had struggled to explain the latest experimental probing of the microscopic realm, crashing atoms into atoms and sending beams of subatomic particles through electric and magnetic fields, with the mathematics of Newton and Maxwell. Even with all its success, this well-established set of theories and scientific laws—now called *classical physics*—could not be contorted to fit the observations the experiments produced.

As all knew then—and we still know now—when we multiply numbers, it does not matter in which order they appear. One times two times three is the same as three times two times one. But this simple and seemingly obvious math did not fit what the new experiments exploring quantum mechanics revealed.

$$A = \begin{bmatrix} 0 & 1 \\ 1 & 0 \end{bmatrix} \quad B = \begin{bmatrix} 1 & 0 \\ 0 & -1 \end{bmatrix}$$

$$A \times B = \begin{bmatrix} 0 & -1 \\ 1 & 0 \end{bmatrix}$$

$$B \times A = \begin{bmatrix} 0 & 1 \\ -1 & 0 \end{bmatrix}$$

Heisenberg's bold idea was to use new, abstract mathematical objects that could be multiplied together but such that the answer depended on the order in which they were multiplied: A times B might not equal B times A. This of course looks strange the first time you see it, but a quick calculation with tables of numbers produces an undeniable proof. Such a table of numbers is called a *matrix*, and Heisenberg's mathematics became known as *matrix mechanics*.[5] It is simply known as quantum mechanics now.

By no means obvious to Heisenberg—or any physicists of the day—was a consequence of this shift in mathematics. The outcome yielded an extraordinary characteristic of quantum mechanics, that we can never precisely know all the properties of an object, something now known as the *uncertainty principle*. This is an excellent illustration of a recurring theme in quantum physics: the mathematics states something that because of our preconceptions about the way the universe functions, we are not ready to accept is true. In this case, Niels Bohr, one of the fathers of quantum physics, suggested that the uncertainty principle implores us to reject the very idea that *things* exist.[6]

When physicists talk about "things," they are usually thinking about a set of properties. A ball, for example, has a shape, a color, a place in space and time. It's these properties that quantum physics, through the uncertainty principle, render undefinable in the quantum world. We just cannot say that a ball has a definite set of properties. This means that there is no experiment that can be performed that can definitively determine the properties of an object, irrespective of the level of precision.

This does not seem to bother us when thinking about the abstract and imperceptible world of quantum particles. However, when we extrapolate the conclusions up to the human scale, our minds start to tangle. As Einstein frustratingly put it, "I like to think the Moon is there even if I am not looking at it." But it's not that the Moon is not there; it's that "there," as a single, precise, well-defined place, is not something quantum physics permits us to define.

In our everyday world and in the large movements of celestial bodies, the uncertainty introduced by Heisenberg is too small to notice. Measuring the mass of a 150-pound person is not going to be affected by whether the measurement varies by the tiny mass of an electron. But at the extremes, in the microscopic world of atoms and electrons, uncertainty and all its consequences reign supreme. And if energy cannot even be defined in the vacuum of space, it may manifest itself with any value. As it cannot be defined, it cannot be predicted and thus will fluctuate.

As Einstein told us, using the most famous equation in the world, $E = mc^2$, energy and mass are directly related. Fluctuations in energy expose themselves as unending sequences of creation and annihilation of particles (mass). We imagine particles spontaneously popping into existence as pairs, a particle and an antiparticle. We'll explore particles and antiparticles more a bit later, but for now, know that the two quickly recombine and destroy each other. Every so often, however, there is an interaction with other particles. It is then that even a modern nonscientist would say, "Things just *got real*."

Physicists often call these quantum fluctuations *virtual* particles, since they are very short lived, existing only fleetingly before vanishing back into the vacuum. But when an interaction interrupts the cycle, a virtual particle can become *real*. This opens up the possibility of a host of interesting phenomena. Perhaps the most interesting—certainly to the question at hand—is the possibility of a universe of particles being born spontaneously out of the vacuum (or the closest thing quantum physics allows to nothing).

Even in the time it takes you to say the word *nothing*, a lot can happen in the timescales as measured by the early universe. The first era of the universe, as we currently understand it, lasted only about 10^{-43} seconds. That is 0.00, followed by forty more zeros, then a 1. That is:

$$0.00000000000000000000000000000000000000000001$$

An unimaginably small chunk of time. What can we compare this to? How can we, as humans, get a sense of such a small scale? Frustratingly, the answer is we can't. This amount of time is much, much smaller than any of our current theories of physics can explore.

But even if we cannot describe in every detail the physics of what happened in that first instant, our current theories can still give us clues. After all, whatever the "correct" theory ends up being, it must still be consistent with our current theories, at least where they apply. Think of a map that depicts a flat Earth. Eventually determining that the Earth was a globe did not instantly invalidate all maps. The smaller the area of a globe you are looking at, the closer reality resembles the map—the two are consistent in this regime. Similarly, Einstein's general theory of relativity becomes Newton's gravity when gravity is weak, and quantum mechanics becomes Newtonian motion when things get large.

So we look to our current theories for guidance. Or, in less principled-sounding words, when you all you have is a hammer, everything looks like a nail. Our hammer is the uncertainty principle, and the nail is the question of creation.

A Universe Born of Nothing

In 1973, physicist Edward P. Tryon published a paper in the journal *Nature* with the title "Is the Universe a Vacuum Fluctuation?"[7] Since then, the idea that the universe was born as a vacuum, or quantum fluctuation, has grown. Perhaps our universe was born from a quantum fluctuation in a preexisting universe, with all the particles and energy bursting out of the darkness. But could our universe, including space and time themselves, be born from a fluctuation in a true nothing?

A meditation session often starts with the task of sitting and *doing* nothing. This is easy enough. But next, the task is to *think of* nothing. Who knew that thinking of nothing would be so difficult? And knowing a little quantum physics does not make it easier. Let us imagine nothing—nothing at all. Physically, we want no space, no time, no energy, and so on. First, "no energy" is too vague an idea, since energy can be positive or negative. Nothing, then, sounds more like zero energy. What do our theories of physics say will happen to this nothing?

The existence of quantum fluctuations, via the uncertainty principle, suggests that our everyday concept of nothing—zero energy, say—is flawed. The thing we are imagining cannot have exact, static, unchanging, and uniform zero energy. Energy, as a quantity, does not have a predetermined value according to quantum mechanics and fluctuates between our measurements. We can, however, define an average value of these fluctuations. With an equal balance between positive and negative fluctuations, this average value can be zero.

OUTCOME							
BALANCE WITH BOOKIE ONE	$1	$0	$1	$0	-$1	-$2	-$1
BALANCE WITH BOOKIE TWO	-$1	$0	-$1	$0	$1	$2	$1
TOTAL	$0	$0	$0	$0	$0	$0	$0

A useful, if simplistic, analogy to understand the statistics of a zero average is a gambler betting at even odds on coin tosses. Here is the strange thing about this gambler: imagine the gambler facing two separate dealers, each flipping a coin. For one dealer, the gambler bets on heads to win on every toss. On the other, the gambler bets on tails to win every time. Pointless, yes, but bear with it. On every toss, the gamblers win one bet and lose the other. Their net gain, or loss, is zero. Over time, the amount owed to this gambler fluctuates but at the same rate as the winnings from the other. What is owed and what is gained always cancel out.

This gambler's coffer is analogous to a zero-energy universe. There is plenty of energy to see in the universe—all those particles of mass and $E = mc^2$ are hard to miss after all. As well as this positive energy, there is also negative energy in the universe. In fact, the gravitational energy stored in the force between masses, the *potential energy*, is negative. This might seem a little strange but is very well defined in physics and just means that we have to input energy to pull two masses apart. If we take all the positive energy and all the negative energy, adding them up over the entire universe, they could cancel each other out, and—voilà!—we get a zero-energy universe.

This idea, of a universe born of quantum fluctuations out of nothing, is relatively new when compared to the myriad of philosophical theories of creation. Before Edward Tryon proposed it in the early 1970s, the consensus was simply that no *scientific* consensus was possible, that the question of what came before the Big Bang was not answerable, at least not by science. It still seems that without considering quantum mechanics, the question is unresolvable. It is only by considering the potential quantum features of a unifying theory that we can propose candidate solutions to the question *why is there something, rather than nothing?*

A Zero-Energy Silver Bullet?

The idea that the universe came from nothing, a true nothing with no space and no time, is rather neat. It leaves no loose ends to tie up! Any question about the origin of the universe will most likely contain the statement "from nothing." Like a frustrated parent shouting "just because" to the endless questioning of a child.

The zero-energy needs of the from-nothing universe add to the neatness. Another thing that doesn't need to be explained is that Heisenberg says the universe can last forever. The from-nothing universe seems like a winner. Everybody's happy! Well, not everybody! While seemingly neat, the from-nothing hypothesis is horribly unsatisfying to many scientists. Common sense, which itself is a terrible guide to understanding the scientific workings of the very small and very large, tells that there must have been a "before" and that something caused the universe to come into being. But at a time before time existed, what does before even mean?

In fact, most cosmologists are not satisfied with the from-nothing universe origin theory, and the search for alternative explanations has continued unabated for several decades.[8] Still, no matter how much they stared at the equations of the general theory of relativity, the solution wasn't there—at least not without making some radical changes to the fundamental makeup of Einstein's ideas. So where could the cosmologists turn? Again, they turned to quantum mechanics.

Perhaps the solution lay not in quantum mechanics offering a birth to the universe in a quantum fluctuation but in reconciling the incompatibility of gravity and the other forces? This remains an unsolved problem, with a "theory of everything" seemingly as far away today as it was decades ago. However, physicists are a clever lot, and there are ways of joining together quantum mechanics and gravity, not in a perfect way but at least in an approximate way. We don't know if the approximation is correct, but perhaps it is, and perhaps our guess is pointing us toward the true theory of everything.

As you might imagine, there are many, many possible theories for how physicists could glue the fundamental forces together, and journal pages are full of the various ideas. However, until we know the mathematics that truly unite the two, there are several ways quantum mechanics could explain the birth of the universe.

Perhaps, in the earliest stages of the cosmos (at least what we currently think of as the earliest stages), the fundamental forces acted nicely together so that gravity didn't completely dominate the others. This is very different from the present-day cosmos where gravity reigns supreme over the large-scale universe, with the other forces dominating only small scales. In the early stages of the life of the universe, perhaps quantum mechanics truly dominated, with gravity being overwhelmed to prevent the indefinite squeezing that would lead to the infinite density and temperature, the *initial singularity*, that dogs the birth of the universe in the standard Big Bang picture.

Without the infinite squeeze, the space and time of our universe could possibly be connected to other structures of space and time, perhaps other universes that came before. Of course, we don't really know how any previous space and time are connected to ours, but there's plenty of room for speculation. Thoughts range from our universe being born from the formation of a black hole in a previous universe to the collision of long-dead universes giving birth to ours in a huge, many-dimensional superspace known as the multiverse.[9] There are many more theories, some much crazier sounding than others, and there will be even more still until we crack the theory of everything.

At this point, we must leave the birth of the universe, because there is so much more to come. We have so far covered the first tiniest fraction of a second of the universe's existence and still have an apparently infinite amount of time ahead until any demise. We must move on, into the future.

Maybe our universe was truly born from nothing, from a quantum fluctuation in a degree of nothingness we struggle to understand. Or maybe quantum mechanics offers a different solution, a way through the infinite density and infinite temperature that existed at the start of the universe. Beyond this point, the *singularity*, there may lie an entire past we have yet to imagine. Now now we step into the next stages of the universe, from a time when the universe was coming into being to a time when it was taking form. We'll see that behind the curtains of the cosmos, quantum mechanics is still playing its vital role.

Why is the universe so smooth?

The universe is immense. Due to the finite speed of light, our telescopes see not only through space but also back through time. They reveal most of the history of the universe, peering all the way back to a few hundred thousand years after the Big Bang.

Why can't we see the Big Bang itself? After the first few minutes of the universe's existence, minutes that saw the formation of the first atomic nuclei, the universe was still extremely hot, with electrons zipping through the soup of matter and radiation. These high-speed electrons were moving too fast to join an atomic nucleus to create the atoms we know today.[1] Instead, the universe was full of *plasma*, with free electrons jostling with light rays, making it opaque. After about 380,000 years of cosmic time, the universe was cool enough and the electrons slow enough for them to stick to the atomic nuclei. In a moment, the universe became transparent.

Once the universe was transparent, light could travel freely across the universe and to our telescopes. But trying to peer more deeply into the universe, into the time when it was opaque, is like trying to stare through a brick wall.

The energy that kept the electrons zipping in the early stages was due to collisions with the immense sea of radiation: high-energy gamma rays, X-rays, and ultraviolet light and radiation left over from the fires of the Big Bang itself. If any electron did manage to grab onto an atomic nucleus, a collision with one of the huge numbers of marauding *photons*, the particles of light, was inevitable and would rip them apart again.

It was the expansion of the universe that cooled these photons from their high energies. As the photons cooled, collisions with the electrons eased, and the electrons became more sluggish. At last, the first real atoms could form. This radiation remains and continues to cool, no longer interacting with atoms but always lurking in the background. We still see this radiation today, but it has now cooled from the extreme temperatures in the Big Bang to a few degrees above absolute zero. Instead of being the highest energy photons, that radiation exists closer to the radio part of the electromagnetic spectrum and is known as *cosmic microwave background radiation*. It is the oldest light we can see.

The immense size of the observable universe is related to the fact that it has had almost fourteen billion years to expand. But as astronomers were beginning to really understand the expansion of the universe, there were some peculiarities that started to bother them. In whichever way they looked, astronomers saw essentially the same stuff: stars and galaxies as far as their telescopes could spy.

Take a telescope in the Northern Hemisphere and point it at a random piece of sky. What do you see? Well, in the nearby universe, you see individual stars in the Milky Way, getting fainter and fainter as you look to larger distances. Then you see other galaxies, large in the sky, as they are also not too far away. More galaxies are apparent, smaller and less formed, because, since light travels at a finite speed, you are seeing them in the past. Eventually, you see tiny galaxies, barely formed, in the early universe, their light having traveled for many billions of years. If your telescope can see into the radio wavelength, you pick up the glow of the cosmic microwave background.

Repeat the experiment with a telescope in the Southern Hemisphere, pointing in the opposite direction from its northern counterpart. What do you see here? Again, there are stars in the Milky Way, similar to the ones you saw in the north but in different patterns and constellations, but that is expected, as we live deep within the galaxy. Beyond our local stars in the Milky Way, you see galaxies, lots of galaxies. Again, not the same ones as seen in the north, but similar in size and shape.

As the deeper universe comes into view, you see less formed galaxies, then, farther away, baby galaxies, and then the impenetrable wall of the cosmic microwave background. You realize that while the details are different, the general view from this telescope in the south is very like the view from the north. Something that appears the same everywhere one looks must be very smooth.

In fact, it doesn't matter which way you point your telescope, the view is generally the same. And this is curious! Why? Because the patches of universe you spy through your telescope can be separated by immense distances, many billions of light-years. These patches should always have been separate, never influencing one another. So why does the distant universe on one side of the sky look so much like the universe on the other side of the sky? Shouldn't they have started slightly differently and evolved differently and thus look very different to us today?

Perhaps everywhere was almost identical at the start, following a similar evolution in all patches? But physicists don't like this idea, as it means that the initial state of the universe was *fine-tuned* to be the same everywhere, and the assumption of such fine-tuning is frowned upon—scientists are very suspicious of special conditions needed to explain experiments and observations. Of course, the birth of the universe could have been special, with the unknown process that brought it into being, a process we do not understand, demanding that it be smooth and identical everywhere.

Is there another way, a physical way, that can smooth out the initial universe? Is there a way that makes everywhere in the universe so similar today?

A Mountain of Energy

The answer is yes, but to understand how, we are going to have to take a bit of a detour. Let's start by imagining a mountain range, with high peaks and deep valleys, ending at the sea. Now imagine placing a ball on one of the peaks. Where will it go? Obviously, it will roll down into a valley. It will tend to lose energy from friction, heating up the ball and mountain a little and eventually settling into the lowest spot it can find. When the ball is higher, upon a peak, it has more *potential energy*. When the ball falls, potential energy is converted to kinetic energy, which eventually is converted into heat. This is a general law of the universe—potential energy will eventually be minimized, and the lost energy will end up as heat. You may have heard of this as the *second law of thermodynamics*.

Sea level, for our purposes, is the point of lowest potential energy. So why in this case is the ball in a valley and not in the ocean? It has more potential energy to lose after all. While the ocean is *the* minimum potential energy, every valley is a *local* minimum of potential energy, at least nearby. Such locations are known as *stable equilibria*. When referring specifically to the ball in its location in a valley, we say that the *state* of the ball is stable; the ball just stays where it is. However, at all other places outside the valley, the state of the ball is unstable, and it begins to roll to somewhere else. To get the ball out of the valley would require enough energy to roll it over the nearest peak. Such a transition from one stable state to the next requires a catalyst, an injection of energy to get the ball moving from its stable location.

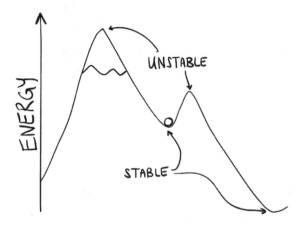

This picture is a powerful one to keep in mind whenever we speak about energy and the ways it is converted from one form to another. We are not actually interested in balls rolling on hills. The ball will represent the thing we are interested in, the mountain represents the potential energy it has, and the location is the *state*, where state simply means a summary of all the information we have about the thing. Everything about this picture is guided by our intuition, hinting that we are talking about classical physics again. But we should be expecting a quantum twist!

It did not take long for mathematicians to seize upon the early success of quantum physics in filling in knowledge gaps left by classical physics. Theoretical advances came to be viewed as modifications of classical theories to fit the new paradigm. This process was referred to as *quantization*. The architects of quantum physics were said to have *quantized* Newton's physics: the laws of motion and how things responded to forces. But Maxwell's equations of electrodynamics had remained intact. To quantize electrodynamics required powerful mathematical tools. Bringing these to bear revealed the possibility of new physical forces—the weak and strong nuclear forces.

These forces are summarized in what is called the standard model.[2] This rather boring name encompasses a powerful mathematical recipe book that allows us to precisely calculate how each of these forces behave. The standard model also describes associated particles, which are the quantized packets of energy mediating each force. Unfortunately, the standard model comes with a lot of jargon, which we will try cover quickly.

To summarize the standard model, there are four force particles that sit under the umbrella of *bosons*. Named after Indian physicist Satyendra Nath Bose, the four bosons are the *photon* of the electromagnetic interaction, the *gluon* of the strong interactions inside atomic nuclei, and the enigmatically named *W* and *Z*, which are responsible for the weak force, a type of radioactivity. Accompanying these is the *Higgs boson*, maybe the most famous of all, which is related to the process by which particles acquire mass.[3]

As well as the force bosons, there are twelve particles that make up matter. These are called *fermions*, named after Italian physicist Enrico Fermi. Six of them are the quarks, which are the only particles that feel the strong force. With the names *up*, *down*, *strange*, *charm*, *top*, and *bottom*, these combine in various ways as the fundamental building blocks of matter, including the more familiar proton and neutron.

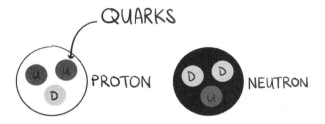

The other six fermions are the leptons, which include the *electron* and its two progressively more massive cousins, the *muon* and the *tauon*, as well as three flavors of a particle known as the *neutrino*. We will be meeting each of these characters in more detail later in the book.

There is one omission from this standard model though: gravity. So far, gravity has eluded quantization, though not for want of effort from the scientific community! Many proposals attempting to integrate gravity into the standard model exist, and some physicists have spent their entire research careers tackling this problem. New quantized theories come and go, most dashed by a lack of experimental evidence to support them.

The experimental playgrounds in the search for new

quantized theories are the exciting research laboratories of theoretical physics—*particle accelerators,* such as the immense Large Hadron Collider at the Conseil Européen pour la Recherche Nucléaire (CERN) on the Swiss–French border. Inside these accelerators, particles are accelerated as close to the speed of light as possible and smashed together. The collisions give rise to explosions of new particles. For the last sixty-odd years, these experiments have given us a huge variety of previously unknown particles. Bigger and bigger colliders continue to be built to search for hypothesized—and maybe even unexpected—particles. Some of these particles are the signatures of fields of energy that underly the universe, the most famous being the Higgs boson.

A potential solution to why the universe looks so smooth is another energy field called the inflaton field.[4] The associated particle, something known as the *inflaton,* is still hypothetical, existing only in the early universe.[5] The fact that we don't see it today suggests that something dramatic must have happened, possibly the most dramatic event in the life of the cosmos. But before we get ahead of ourselves, let's return to our mountain of energy.

On the classical energy mountain, valleys are places of stability. A state is either stable or unstable—it is black or white. On the *quantum* energy mountain, stability exists on a spectrum of gray. Looking back at our mountain of energy picture, the sea level was the lowest potential energy. In quantum physics, this is called the *vacuum state.* The kinetic energy of the ball—when it moves away from a valley—has its analog in quantum excitations and manifests as particles. So the energy associated with a particular type of force is in a valley when no particles are present and moving along a hill when particles are present. A ball can be trapped in a valley and stay there forever, never reaching the sea—its vacuum state. But in quantum physics, where motions and positions are uncertain, things are more interesting. A spontaneous quantum fluctuation might be just the catalyst needed to create what is called a *phase transition*.

Water Break

Now it sounds like things are getting quite complicated with fancy quantum physics jargon. What is this "phase transition," for example? The answer lies in a glass of water. Imagine taking some part of the water and replacing it with another part of water. Would it appear to be the same cup of water? Probably. Water appears pretty uniform. It has a property to it that physicists call *symmetry*. But let that water freeze before adding the new part, and something different emerges. In fact, every time we freeze the water, the blocks of ice that result look subtly different.

If you have a freezer at home with a tray of ice cubes, take some out and examine them. Each ice cube has a different pattern of trapped air bubbles, cracks, and other defects in it. Replacing one part of the ice cube with a piece from another ice cube results in a visibly different cube of ice. The ice, we say, is not as symmetric as the water that it was frozen from. Heating water instead of cooling it has the opposite effect. Gaseous water vapor is even more uniform than liquid water. Generally speaking, the hotter something gets, the more uniform it is. The reason has to do with how densely packed the energy is.

You may recall learning about these familiar things in elementary school as *phases* of matter. Each compound, like H_2O for example, can exist in a solid, liquid, or gas phase. For H_2O, these states are ice, water, and water vapor. These are called *classical* phases of matter. Including quantum physics in the mix results in dozens more, aptly named *exotic phases of matter*.[6] Quantum states and their phases are not so easy to visualize, but when they change from one to another, the results can be just as rapid as water molecules escaping from liquid into gas or being locked into place to form a rigid crystal of ice.

Going back to our trusty mountain of energy, the sea level is like very cold ice—the lowest energy classical phase H_2O can take. There is more energy in liquid water, which is analogous to a high valley. Higher still at the mountain peak is comparable to water vapor. When we heat ice, we get it over the peak and into the liquid valley. More heat gets it over the next peak into the gas valley.

Going the other direction is less obvious. Say we start with liquid water. Keeping it at a fixed temperature above 0°C (32°F) keeps it happily bouncing around in its valley. Now, lower the temperature of the surroundings. Energy is lost, but this only means the water settles lower and lower in its liquid valley. How does it ever make it over the peak to the solid plains—sea level?

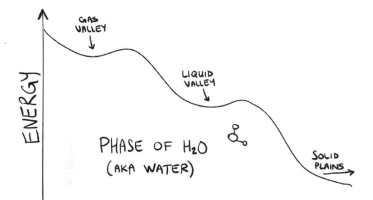

The short answer is, it doesn't. Not without help anyway. This is something else that can be tried at home. Take a bottle of purified water and place it in the freezer. We would expect that as it crosses 0°C, it will turn to ice, and this is what happens in normal matter with impurities, as these provide the sites for freezing to start. But with pure water, without impurities, the water doesn't freeze! You will find in the freezer liquid water at −18°C (−0.4°F) (a typical freezer temperature). In fact, if you were careful enough, you could create liquid water at a temperature all the way down to about −50°C (−58°F)! This is *supercooled* water trapped in the liquid valley.[7] Water with impurities has an energy landscape with less pronounced valleys, and cooling the water slides it down the mountain without it getting trapped in any valleys.

At −18°C, giving supercooled water the slightest asymmetry, say by hitting the side of the bottle, sets off a chain reaction as the water tumbles down the potential energy mountain, releasing energy to its surroundings as it settles into a new minimum.

Now stretching our imagination into the abstract and hypothetical, suppose there was something much less tangible than the thermal energy of water—a new type of energy that couples to space and time in such a way as to cause space itself to expand. In a high-energy state, this would cause space to expand rapidly. This expansion is now called *inflation*. This state is analogous to supercooled water—it's in a high-energy inflation valley. Much like the freezing of supercooled water—perhaps due to a quantum fluctuation—a phase transition occurs, and we make it out of the inflation valley, falling down the mountain toward the vacuum. As we do so, particles are created—the inflatons.

As mentioned before, in the universe of today, general relativity dominates the large and quantum physics the small. But at this point in the early universe, the scales are tipped. The picture we have created is one of quantum physics controlling the cosmos at its largest scale, and this results in an immense cosmological event! By combining quantum physics and general relativity in this way, the equations reveal an unfathomably rapid expansion, with every patch of space expanding much, much faster than the speed of light in the inflationary state.

While an extreme event to say the least, inflation does quite nicely explain why the universe appears to be so uniform. Before inflation, the universe was immensely hot and dense and probably a complete mess of conditions that varied from place to place, even between extremely small distances. Superimposed on this boiling sea are the tiny fluctuations demanded by quantum uncertainty. Inflation occurs, spreading out the energy of the universe in all directions, with a tiny patch spread out to encompass our entire observable universe and much beyond. The result is that in our universe, the density of energy is the same everywhere.

Inflation is a pretty compelling theory, and if you pick up any modern textbook on cosmology, a discussion of the inflating universe will be in there somewhere. Like a detective in an Agatha Christie murder mystery, inflation seems to tie up many loose ends and solve tricky questions about the universe we see around us. Surely, it is one of the great successes of modern cosmological thinking?

A Dark and Mysterious Matter

While inflation elegantly appears to explain the observed universe, it is not a complete theory, as there are a few loose ends that need to be tied up to seal the deal. First, there is the question of the nature of the inflaton. Where did it come from, and where did it go? And does the inflaton have a role in the universe today? Some scientists think that the inflaton has morphed into another cosmological force, namely *dark energy*, a force we will discuss later, but at the moment, this theory is quite speculative.[8]

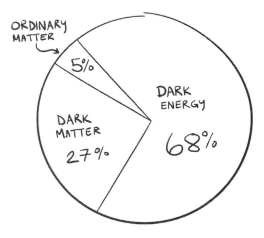

One challenge in confirming this theory is the lack of direct observational evidence for the period of inflation. You might be scratching your head over this, thinking "We started this story by asking the question of why the universe appears to be identical in all directions. Isn't this direct evidence for inflation?"

Yes, it is, but it is not conclusive evidence, as the universe could have simply been born nice and smooth and the same in every location. The same is true for the other hypotheses relying on the idea of inflation, from the *missing monopole problem* to the *flatness problem*, ideas that we don't have time to go into detail about now but that you could spend many hours googling. In fact, all the observations of the universe that can be used as evidence for inflation are also completely consistent with the fact that the universe could have been simply born this way.

When you have competing ideas in science that can explain the same observations equally well, scientists have to scratch their heads a little. First, they have to ask themselves, "which competing idea is more likely?" For inflation to be true, we have to invoke a completely unknown force, the inflaton, that appears, radically changes the universe, and disappears in an instant. This might be related to the complex physics underway in the early universe as it cooled from tremendously high temperatures, but again, some of these ideas are more speculation than science.

Alternatively, for the "the universe was just born this way" option, we have to imagine that the process that brought the universe into being did so just right to ensure that everywhere was exactly the same: precisely the same density and temperature, precisely the same constituents of particles and radiation, precisely the same expansion, uniform in every direction. As we hinted at before, scientists really don't like this kind of fine-tuning of the universe, as small differences or fluctuations in properties between locations seem inevitable. But in truth, we have no real idea how the universe came into being or about the physical mechanism that laid out the properties of the universe, so maybe cosmological birth is the one place where perfection has been achieved.

How can scientists decide which theory is correct? What you need is more observational evidence, evidence that can discriminate between the two ideas. Astronomers are on the hunt for such a telltale signature, waves of gravity imprinted on the universe, to distinguish inflation as the reason why our universe is so similar. If inflation is indeed the correct description of the very early universe, the evidence might be conclusive in only a few years.

But before we close, remember those small quantum fluctuations that existed before inflation? They, too, must have blown up during the period of rapid expansion. If the theory of inflation is correct, they were written into the matter distribution of the postinflation universe, ripples in density of an otherwise smooth cosmos. It is these small differences in density, seeds at one part in ten thousand, that allowed gravity to do its work and matter to pool into the galaxies, stars, and planets that we see around us. Without these seeds, none of this—you, me, the Earth, the Sun, or the Milky Way—would exist. We owe our existence to the action of the quantum.

Why is there matter in the universe?

Obviously, there is matter in the universe. A lot of matter! Matter locked up in stars, planets, and rocks sprinkled in between. An immense amount of matter is distributed as gas between the stars and the galaxies, spread throughout the universe. There is matter as far as we can see. But why is it there? Why is there any matter at all in the universe?

This might seem like a frivolous question. Surely, this is just obvious! If there was no matter in the universe, we wouldn't be here to ask the question. But in terms of our understanding of the fundamental makeup of the universe, the existence of matter presents us with an immense challenge. To understand why it exists, we need to think about the conditions in the universe just after the burst of expansion during inflation.

As inflation ended, its energy was dumped back into the

universe, into the particles and radiation that provide the basic building blocks of all the matter around us. But the temperatures were so hot that the normal, everyday matter we're familiar with didn't exist. Only the fundamental building blocks existed: the quarks, the electrons, and the superhot photons. Our laws of physics, as far as we know them, tell us that this soup was an equal mixture of matter and antimatter. Electrons were accompanied by their positively charged antimatter siblings, the positrons.

The existence of antimatter was predicted in the 1920s by theoretical physicist Paul Dirac.[1] Dirac was actually attempting to unite quantum mechanics with Einstein's special relativity to understand the properties of the electron. His equations, however, threw up two solutions, one negatively charged, which he knew represented the electron, and an identical but positively charged sibling. He was not sure what to make of this, wondering if he had managed to accidentally write the much heavier proton, the most common positively charged particle, into his equations. Shortly after, positively charged electrons, now known as positrons, were detected in experiments, and the scientific community realized that every particle of matter possessed an antimatter twin.

In the early universe, other particles were in the mix, the funnily named *quarks* and their antiparticle equivalents.[2] Like electrons, quarks are fundamental particles, meaning that we cannot chop them into smaller pieces, but they are less famous than their cousins, the electrons. This is because unlike electrons, which can be found on their own, quarks are always bound up in other particles, being the primary constituents of the protons and neutrons that form the nuclei of atoms.

Interesting things can happen in such a hot soup of fundamental particles. Electrons can collide with positrons and be completely annihilated, instead creating two photons of radiation. The same is true for quarks that encounter antiquarks: annihilation and the creation of more photons.

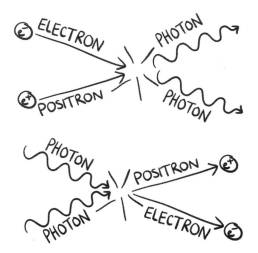

The reverse is also true: two colliding photons can make an electron-positron pair or a quark-antiquark pair. As long as there is a lot of energy, the situation remains in equilibrium, with as many electron-positron pairs being annihilated and turned into photons as pairs of photons into electrons and positrons.

Don't forget that at this point, the universe was still expanding and cooling as it expanded. The expansion continually robbed the photons of their energy, with their wavelengths getting longer as the universe aged. What happened to all the matter in the universe?

Things started to get interesting when the universe was about 10–11 seconds old, long after the period of inflation had ended. The end of inflation flooded the universe with energy, a mix of matter and antimatter in a soup of high-energy radiation. But by this point, the photons in this superhot soup no longer had enough energy to create particles when they collided, so the universe became unbalanced. No more electron-positron pairs were created, and no more quark-antiquark pairs were produced. There were still particles in the mix, both matter and antimatter, and these could still collide, be annihilated, and create photons. Very rapidly, all the electrons met up with positrons, and in an instant, they transformed into photons. The same is true of the quark-antiquark pairs, rapidly being annihilated and turning into more photons. So once the universe passed this critical cooling point, all the matter had turned into radiation, and there should have been no particles left in the universe. After this point, the universe should have had no more matter.

This is clearly not the universe we inhabit. In our universe, matter dominates, and antimatter seems to be extremely rare. Antimatter is sometimes spat out of radioactive materials, created in particle physics experiments, or seen to be emitted from some of the more exotic processes in the universe. But matter dominates the universe.

We've already mentioned the cosmic microwave background, the leftover radiation from these early times in the universe. This radiation must have come from the particles and antiparticles being annihilated. If we count the number of photons in the cosmic microwave background, there are about a billion for every one of the pieces of matter, the protons and neutrons found in the nuclei of all atoms.

This seems to suggest that somehow the universe was already unbalanced before the final annihilations took place—it was not, in fact, a perfectly even mix of matter and antimatter to cancel itself out. For every billion positrons in the universe, there must have been a billion plus one electrons, so that after the final annihilations and creations, we were left with only electrons and photons in the universe. The same must have been true for the quarks and the antiquarks, with unbalanced annihilations and creations leaving only quarks and more photons behind.

This is quite strange, as our laws of physics appear to be identical or symmetrical for matter and antimatter, with no hints that either one should be more prevalent. The existence of matter today, with no antimatter, tells us this cannot be correct. A break in the symmetry is needed somewhere, but where? Can this really answer the question of why there is any matter in the universe today?

The Mathematics of Beauty

We need to take a deeper look at symmetry. The ancient Greeks, such as Pythagoras and Plato, saw symmetrical shapes as embodying the beauty of nature. Indeed, Aristotle proposed that the heavens were constructed from concentric spheres because the sphere was the most symmetrical and hence the most beautiful shape. Of course, symmetry shows up in many historical contexts: wheels were made to be round, sports balls were spherical, tools and weapons needed to be balanced, and so on.

The intellectual concept of symmetry is thought to have emerged as a stark aesthetic shift coinciding with other Renaissance values, such as simplicity of forms. A wild rose is an ugly chaotic mess of shapes, but a rose painted on the wall of a Renaissance cathedral would be equal in its proportions and pure in its form—in other words, symmetric.

Mathematicians took hold of the concept and refined it over several centuries. As with all mathematical concepts, over time, it became more and more abstract. The mathematical understanding of symmetry started with specific things, like regular geometric shapes, and by the nineteenth century had evolved into the theory of *groups*. A group, roughly speaking, is any collection of things that can be combined to make another in the same collection. Numbers are a perfect example: take two numbers and combine them to get another.

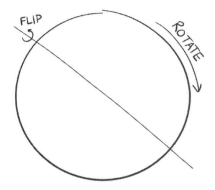

How are groups connected to symmetry? Consider a circle. What can you do to a circle and still end up with the same circle? You could flip it, or you could turn it, but you can't squish it, for example, as that would make an ellipse. The *transformations* that preserve the circle are its symmetries, and they always form a group. Mathematicians figured out a lot of things about groups, so by the time modern physics had been established, physicists could borrow their ideas and make rapid progress. Indeed, symmetry is so important that some have gone as far as to suggest that physics is nothing more than the study of symmetry and that modern physics is applied group theory! And while symmetry remains an intuitively useful concept, the abstract nature of mathematical groups became a crucial prerequisite for understanding the universe—for that is the language of quantum physics.

Here is a general rule of thumb: if you want to discover

something in physics, use symmetry.[3] There are two ways to discover things in theoretical physics, the branch of physics that uses mathematics rather than laboratories to study nature. The first is to look at existing laws and equations and find new symmetries in them that others haven't noticed. The second is to propose new theories that posit symmetry at the outset. Examples abound for each.

Many classical theories of physics possess symmetry. Indeed, Johannes Kepler's laws of planetary motion—which some suggest paved the way for the revolution that led to modern science—have beautiful geometric simplicity. They demand that planets orbit the Sun, tracing the geometric shape of an ellipse. However, Kepler was driven by the need to fit the observational data rather than by the requirement to have symmetry in his laws. In fact, it was 250 years later that German physicists Carl Runge and Wilhelm Lenz were credited with "discovering" all the detailed mathematical symmetries in planetary motion.

Fast forward to 1905, and we again meet Albert Einstein in his miraculous year. He is probably the only person in history so easily associated with a mathematical equation, $E = mc^2$. This time, however, the equation is a direct consequence of a mathematical symmetry. Einstein changed physics by creating a theory from principles of symmetry rather than trying to find equations to fit observational data.

The symmetry in Einstein's general theory of relativity is a symmetry of viewpoints. He thought about someone in a spacecraft, far from any sources of gravity. Inside the spacecraft, everything is weightless, floating in the air, a very different situation from being on the surface of the Earth, where gravity pulls everything down. But then Einstein wondered about someone falling because of the gravitational pull of the Earth. Not just a person but a person inside a room that is falling too. Within this falling room, our person would also see things floating in the air as if there were no gravitational pull at all. Their view would be identical to that of the person in deep space. To a falling person on Earth, gravity no longer exists. Strange as it may sound, this was the foundation of our modern theory of gravity.

While matter may have existed more or less as it does today since the first few moments after the Big Bang, our understanding of it is relatively new. Of course, the ancients were aware of matter and had vaguely accurate conceptions of the elements as we know them today. With the arrival of quantum mechanics and modern atomic theory, we were able to unravel the structure of atoms and reveal that all the elements were built from a small number of fundamental particles. So the question of why matter exists at all is one that is potentially answerable only in the context of quantum physics. Even then, it was not evident in its current form until 1928, when Paul Dirac wrote down his namesake equation.

By the late 1920s, quantum mechanics as a discipline was not yet fully formed. Heisenberg had introduced his early version of quantum mechanics, known as *matrix mechanics*, but it provided as much confusion as clarity: the mathematical language was unfamiliar to many physicists, and it was far from clear what the physical foundations of the theory were. Much of the theory was still fractured. What's more, it was a quantized theory of classical physics and did not account for the influence of electromagnetic force. Dirac was determined to modify the equations to be consistent with the principles introduced by Einstein such that they would possess the appropriate symmetries. The Dirac equation is now considered the genesis of the *standard model of particle physics*, the quantum mechanical description of particles and forces (except gravity!). Though it took many decades to complete in the end, some surprising consequences were immediately evident from this work.

The Dirac equation contains the first scientific prediction of something never before seen in nature. While Dirac did not consider symmetry when developing his equation, the equation itself possesses a symmetry about *charge*. Electric charge is a basic property of matter that allows it to be influenced by electric and magnetic forces. By convention, we consider amounts of charge in units of quantized amount, known as e. Every electron has charge $-e$, and each proton has charge $+e$. If, by magic or imagination, we were to change the charge of an electron from $-e$ to $+e$, we would get a proton, right? No! There are many other differences between electrons and protons beyond charge (for example, the proton is almost two thousand times more massive).

Upping the Anti

Changing the charge of an electron from negative to positive is like flipping the circle we talked about before. But this time, we would expect things to be quite different—after the flip, we would not see the same circle. However, as far as the Dirac equation is concerned, something equal in every way to an electron but with positive charge $+e$ is a valid solution to the equation. In other words, the Dirac equation predicted a new type of matter, *antimatter*. Dirac made his prediction of the antielectron, now named the positron, in 1928. Just four years later, Carl Anderson discovered conclusive evidence for its existence in an experiment where he was studying the impact of particles from outer space, known as cosmic rays.

Antimatter clearly exists and obeys the same laws of physics as matter. Some scientists have wondered if entire galaxies could be made of antimatter! But here is the problem. When matter and antimatter meet, they annihilate each other and produce massive amounts of energy, such as gamma radiation. Antimatter is the ideal fuel for science fiction, perfectly annihilating matter to produce energy and propel future spaceships. But it is difficult to store antimatter. As soon as it touched the vessel you plan to store it in, the two would be annihilated in a massive burst of energy. Similarly, if large chunks of the universe were antimatter, where it met normal matter, it would glow brightly in gamma rays. We don't have any evidence of this happening, so there are probably no immense regions of the universe made of antimatter.

Because of symmetry, our laws of physics would work just the same if we flipped the charge of every particle. Physics on paper doesn't discriminate between matter and antimatter, so why does nature? Antimatter was born out of quantum symmetry, but we need something else to explain why there is more matter than antimatter today. We need to break the symmetry that created it, which means we need to find some asymmetry in the laws of existing physics, or we need new physics that allows for matter/antimatter asymmetry.

But why do we build up symmetries only to call them broken later? Why not just go with the asymmetric description in the first place? The answer is attributed to Emmy Noether. Noether was a prominent mathematician who, like Einstein, made contributions to many areas of physics.[4] She proved perhaps the most important theorem about symmetry, which states that every symmetry corresponds to a conservation law.

To physicists, conservation laws are about as sacred as it gets. They are extremely powerful tools that drive the intuition behind most of our understanding of the universe. In the case of the circle and its rotational symmetry, for example, Noether's theorem implies that rotating objects will have a conserved quantity related to their spin, something we know as *angular momentum*. Symmetries are sought by scientists when looking at existing theories and creating new ones because they are both beautiful and simplifying—a circle requires only one number to specify it (the diameter), for example. Nature possesses symmetries in many places. It behooves us to find those symmetries, because knowing them allows us to create economical descriptions of the physical world in terms of simple conservation laws. But in other places, such as in the case of matter and antimatter, nature is asymmetrical. We don't really know when and where we will find symmetry—they are usually true eureka moments—and perhaps that is what gives scientists the thrill of discovery.

Wherever a symmetry fails, a conservation law is broken. And as every fictional police officer has said, without law, there is only chaos. This isn't quite right, of course. We happily live in a world, both physical and social, that is not *too* constrained by law. A completely symmetric world would have nothing interesting to say about it. Such was the state of the universe at the first instant of the Big Bang. So the question of where all the antimatter went is a very important one: *By what physical process did this symmetry get broken?*

Deep within the details of the standard model, charge symmetry can be broken. Though we now have experimental evidence of nature's ever-so-slight preference for matter, it is not enough asymmetry to account for the discrepancy between matter and antimatter. We still need a symmetry-breaking mechanism. Most proposals posit new models of phase transitions as likely explanations, much like those we discussed before between water and ice. Others argue that physics beyond the standard model is needed.

The lesson so far is that for matter to exist in the universe, it appears that the laws of the universe must be fractured. In a perfect universe, with perfect, symmetrical physical laws, there would be a particle for every antiparticle. Their annihilation would be perfect, leaving behind only a sea of formless radiation, with no matter to prove the existence of an asymmetry.

While the details about it are still uncertain, scientists are sure this imperfection lies within the laws of physics. Today, it is mostly hidden, rearing its head so rarely as to be invisible, but in the earliest stages of the universe, with so many collisions and interactions underway, this imperfection must have played its part in a major way, ensuring matter outnumbered antimatter by one part in a billion.

Such asymmetries were not only at play in the early universe but are also present today, with the results of particle physics experiments showing signs of imperfections: symmetries that almost hold but not quite. This means that quantities we thought were conserved in the universe are really not! We will meet the ghostly *neutrino*, a particle that barely interacts with any other particles, a little later, but experiments have shown that it violates a fundamental law of the universe, something called *parity*.

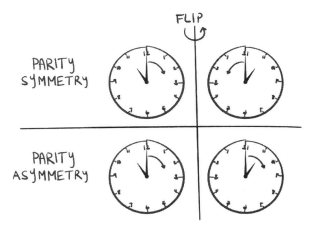

To understand parity, think of watching an old-time movie in an old-time cinema. How do you know if you are watching the intended film or a version that has been mirror flipped by a clumsy projectionist? If you are watching a human drama, the clues become obvious as you notice the number of right- and left-handed people or see writing upon the page. Human civilization has an inherent asymmetry to it.

But what if you were watching a natural scene, of whales crashing through the ocean or an eagle soaring above the mountains? Now the situation is much harder to discern from its mirror image. Maybe if you were an expert on whales or eagles, you might pick up on the visual clues, but the scenes of mountains and oceans look basically the same viewed correctly or mirror imaged.

This appears to be true about the laws of physics as well. An interaction, such as an electron bouncing off another electron due to the electromagnetic force, appears physically unchanged in both its normal and mirror-image view. The same appears to be true for gravity and the nuclear strong force, although interestingly not the last of the forces, the weak force.

Breaking the Quantum Law

The neutrino particle is ghostlike, with no charge and virtually no mass. Its interactions with other matter can only be measured through gravity and the weak force. In the 1950s, scientists realized that reactions involving neutrinos and the weak force refused to conform with this mirror-image requirement. In 1956, physicists Tsung-Dao Lee and Chen Ning Yang suggested that neutrino reactions seen in a mirror simply did not occur in the actual universe. Soon after, experimental physicist Chien-Shiung Wu showed this was true in an experiment looking at the radioactive decay of cobalt, which spat out an electron in one direction and a neutrino in the opposite direction.[5] The fleeing neutrino vanished from the experiment, but Wu could detect the electron. If parity had been conserved, Wu expected electrons from lots of cobalt atoms to spray equally in either direction, but her experiments only detected electrons heading one way. In our universe, parity was clearly broken! This came as a complete shock to the scientific community, which expected the universe and the universe in a mirror to be completely symmetric. A small variation had to be introduced to the mathematics to account for it.

The symmetry of other interactions, such as gravity, electromagnetism, and the strong force deep inside atoms, remained intact, with perfect mathematical symmetry. The conservation of electric charge in electromagnetic interactions is a prime example. We have never observed an interaction that changes the net amount of charge. Scientists have looked, but there appear to be no cracks in its physics.

We don't understand why some physical laws are perfectly symmetrical while others are asymmetrical. We also don't know what governs the scale of the breaking of symmetry and why matter outnumbered antimatter at a level of one part in a billion in the early universe. Why wasn't it one part in one hundred, or one in a hundred trillion? The resultant universe would have been radically different, with much more or much less matter than we currently see. This is something interesting to think about, and there would at least be matter in some form. Without cosmic imperfection, we would not be here to ponder at all.

Where did the elements come from?

The chemical elements are the building blocks of the universe. There are ninety-two natural elements, plus a couple dozen other superheavy elements created in our laboratories. The typical human is about 70 percent water, and we know that water is composed of an uncountable number of identical molecules, each consisting of two hydrogen atoms and one oxygen atom bound together by the electromagnetic force. But many more elements are needed to build a human, with carbon, sulfur, phosphorus, and so on bound together in a myriad of different molecular structures. As we've seen, the early universe was essentially a hot soup of fundamental particles—quarks, electrons, and photons—so where did all the elements necessary to build a human come from?

Let's go back to the early stages of the universe, just a millionth of a second after the beginning, while the temperatures were still extremely high. Eventually, the conditions were cool enough for quarks to combine. There are different kinds of quarks, six in total, each with slightly different properties. Physicists have given them interesting names, including *strange, charm, top,* and *bottom,* but for normal matter in the universe, including the stuff from which we are made, it is the two lightest types of quark, *up* and *down,* that matter. To make a proton, you take two up quarks and a down quark and stick them together. Two down quarks and an up quark form a neutron. The sticking is provided by the *strong force.*

The strong force plays a key role in the discussion of the elements, so let's explore it a bit further. The modern concept of an atom was born in 1911, when Ernest Rutherford showed that all the positive charge of an electron was locked away in a tiny atomic nucleus.[1] This nucleus was a thousand times smaller than the orbits of the electrons, and most of an atom is empty space!

An atomic nucleus is composed of a mix of two kinds of *nucleons*: the electrically neutral neutrons and positively charged protons. Packing protons into the tiny volume of the nucleus means that the electromagnetic repulsion between them is immense. What stops atomic nuclei from blowing themselves apart? The answer is a much stronger force, one that can completely overwhelm the effects of electromagnetism, a force called (unimaginatively) the *strong force* by physicists.

The strong force is fairly complex. Within the standard model of particle physics, the recipe book for the actions of fundamental particles and forces, the strong force exists not simply between protons and neutrons but between the quarks that comprise them. Each quark experiences the strong force through the exchange of another particle, called a *gluon*, which glues the quarks together. Inside each proton and neutron, three quarks are vigorously exchanging gluons, tightly binding them together.

So how is the strong force responsible for binding the nuclei together? When protons and neutrons get close enough, the quarks in one can feel the presence of the quarks of another, and a gluon can be exchanged. Effectively, the strong force binding the nucleus together is the remnant that leaks between the quarks in the protons and neutrons. That's how strong the strong force is!

This means that the strong force between protons and neutrons operates only over a very short range, and these particles have to get very close to feel the force. This need is where problems in creating the elements in the early universe begin.[2] It is true that the high temperatures in the early universe meant that protons and neutrons were undergoing many violent collisions, bringing them close enough for the strong force to latch them together. A proton and neutron could join together to form a *deuterium* (or heavy hydrogen) nucleus, but deuterium is a very fragile nucleus, and in the hurly-burly of the fires of the Big Bang, it was rapidly ripped apart. Without forming deuterium, heavier elements were unable to be forged, a barrier known as the *deuterium bottleneck*.

Eventually, the universe cooled enough for deuterium to

survive the collisions and thus be used as the building blocks for larger nuclei. Two deuterium nuclei could bind together to form the nucleus of a helium-4 atom. If a deuterium nucleus could snare a single proton, a helium-3 nucleus was formed. With that, we appeared to be on our way to building all the chemical elements. However, with the universe continuing to cool, a further hurdle became apparent.

Deuterium nuclei are positively charged and therefore repel one another. With the universe cooled, the motions of the deuterium nuclei slowed. They became sluggish. As they approached one another, the electromagnetic force built and forced them apart. They simply couldn't get close enough for the strong force to reach out and bind them. Free protons were also forced away. After a few minutes in which some helium and lithium nuclei were formed, this *nucleosynthesis* appeared to be over. The pathway to forging heavier elements in the Big Bang was completely cut off. So we are left to ask: *Where did the chemical elements come from?*

A LEGO Universe

Could the deuterium bottleneck have been avoided? Surely, there must be other ways of forging heavier elements. What if, in addition to combining a proton and neutron to create deuterium, we consider sticking two protons (a *diproton*) or two neutrons (a *dineutron*) together and building up elements from there? Wouldn't it be nice to build a universe like a LEGO tower by combining blocks any which way? Alas, this is not possible, as nuclear physics is science, not alchemy. Some reactions are rare, and some are not possible at all.

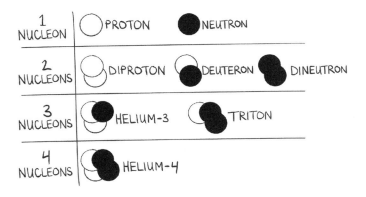

Helium-3 has three nucleons (two protons and one neutron). Helium-4 has four nucleons (two protons and two neutrons). And up we go to uranium-238, the heaviest natural element. To build up elements from protons and neutrons, we have to stick smaller nuclei together to make heavier nuclei. If the starting point is just a sea of individual protons and neutrons, to get started, we have to go through something with just two nucleons. There are only three options for the pair: proton-proton (*diproton*), neutron-neutron (*dineutron*), or proton-neutron (*deuteron*). Naively, we might guess that diprotons are ruled out by electrostatic repulsion—like charges repel after all, and each proton has a positive electric charge. However, the strong force—which binds nucleons together—is named so for a reason. At the scale we are examining here, the repulsive force of the charges is negligible. This is how massive nuclei with many protons avoid the electromagnetic forces. More on that later!

So diprotons, dineutrons, and deuterons all seem like perfectly acceptable LEGO blocks of matter to build upon. But we are missing one thing: *spin*. The concept of spin in quantum mechanics was introduced by Wolfgang Pauli in 1924.[3] He defined it as "two-valuedness not describable classically." Two-valuedness simply means something that can take on two different values (like a light switch), but it has no counterpart in classical physics. How is it best described then? You guessed it—quantum mechanically!

Spin is an internal degree of freedom of fundamental particles. This is why there is no good classical analog. It is one of the first concepts students encounter in quantum physics, usually in a chemistry class. On the wall of every high school chemistry laboratory is the periodic table of the elements. From H for hydrogen to Og for oganesson, they are numbered 1 to 118, yet they appear to be arranged in an odd way. Hydrogen and helium are all alone in the top row, and it starts to fill in as we go down. The reason for this is mostly the way the electrons are arranged in the atoms of each element. The periodic table presents neutral atoms, with their full accompaniment of electrons, but remember that in the early universe, there was only hydrogen and helium, and for several hundred thousand years, it was still too hot for electrons to latch onto the nuclei. But let's continue with the periodic table.

1 H																		2 He
3 Li	4 Be											5 B	6 C	7 N	8 O	9 F	10 Ne	
11 Na	12 Mg											13 Al	14 Si	15 P	16 S	17 Cl	18 Ar	
19 K	20 Ca	21 Sc	22 Ti	23 V	24 Cr	25 Mn	26 Fe	27 Co	28 Ni	29 Cu	30 Zn	31 Ga	32 Ge	33 As	34 Se	35 Br	36 Kr	
37 Rb	38 Sr	39 Y	40 Zr	41 Nb	42 Mo	43 Tc	44 Ru	45 Rh	46 Pd	47 Ag	48 Cd	49 In	50 Sn	51 Sb	52 Te	53 I	54 Xe	
55 Cs	56 Ba		72 Hf	73 Ta	74 W	75 Re	76 Os	77 Ir	78 Pt	79 Au	80 Hg	81 Tl	82 Pb	83 Bi	84 Po	85 At	86 Rn	
87 Fr	88 Ra		104 Rf	105 Db	106 Sg	107 Bh	108 Hs	109 Mt	110 Ds	111 Rg	112 Cn	113 Nh	114 Fl	115 Mc	116 Lv	117 Ts	118 Og	

57 La	58 Ce	59 Pr	60 Nd	61 Pm	62 Sm	63 Eu	64 Gd	65 Tb	66 Dy	67 Ho	68 Er	69 Tm	70 Yb	71 Lu
89 Ac	90 Th	91 Pa	92 U	93 Np	94 Pu	95 Am	96 Cm	97 Bk	98 Cf	99 Es	100 Fm	101 Md	102 No	103 Lr

Some of the concepts used when discussing atoms in chemistry include orbitals, shells, and quantum numbers, which define the different properties of electrons zipping about the nucleus. Pauli invented one of these so-called quantum numbers to explain the arrangement of electrons in the shells of observed atoms. The defined rule is that no two electrons can have the same quantum numbers—the so-called *Pauli exclusion principle*. Perhaps you remember "filling" orbitals in with electrons in a chemistry class—1s, 2s, 2p...3d, and so on.

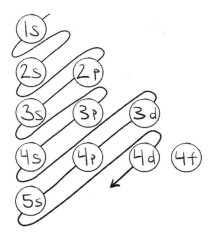

We say all things have spin, with the smallest amount of spin being no spin at all, so a spin of 0. It turns out that Pauli got one thing wrong about spin—the two-valuedness of electrons is not universal. We can have more values, and the allowed values of spin come in halves. So the spin of a given object can be 0, ½, 1, 1½, 2, and so on. Now we understand the fundamental particles with integer spins (0, 1, 2, etc.) behave very differently from those with half-integer spins (½, 1½, 2½, etc.). The former are *bosons* and the latter are *fermions*. The key difference lies squarely in the Pauli exclusion principle, which states that no two fermions can occupy the same quantum state—that is, given a precise description of a fermion (like a proton or electron, for example), no other fermion can have that description. In high school chemistry, recall that you had to give each electron a different quantum number. So once all the internal degrees of freedom—like spin—are used up, fermions cannot occupy the same region of space. Bosons, on the other hand, can bunch up all they like because they do not obey the Pauli exclusion principle. In a sense, this is why matter, built of fermions instead of bosons, takes up space in the first place. Don't blame the turkey dinners over the holidays for the extra inches you gain around your waist—thank quantum physics!

Again, thinking about filling orbitals, the spin of electrons was represented as an arrow—either up or down. That's exactly the two-valuedness Pauli spoke about. Every electron has spin $\frac{1}{2}$, as do the nucleons: the proton and the neutron. In quantum physics, the spin can take on either sign at its extreme. In the case of the electron, this is $\frac{1}{2}$ or $-\frac{1}{2}$. The label doesn't much matter since, as we alluded to, spin is an abstract internal degree of freedom. So $\frac{1}{2}$ or $-\frac{1}{2}$ is just as good as $\uparrow$ or $\downarrow$ is just as good as 0 or 1 is just as good as ☺ or ☹ —you get the picture. The important point is that two fermions with spin $\frac{1}{2}$ cannot have the same spin direction if they occupy the same space, as they would if they were bound together.

So in a diproton, the spins of the protons must be opposite. Otherwise, they would occupy the same quantum state and violate the Pauli exclusion principle. The same is true for a dineutron. However, in a deuteron, the proton and neutron can have the same value of spin since they are distinguishable by other means—their masses, for example.

The Building Block That Wouldn't Fit

So spins, nucleons, exclusion, phew! Time for a recap. It is the early universe. The temperature has cooled enough so that protons and neutrons are free. It's time for them to combine. The options are diproton or dineutron with opposite spin nucleons, or deuteron, whose spins can have any alignment. Now here is the key point for the strong force leaking out of the nucleons: it likes spin. When the spins of two nucleons are opposite, their spins add up to 0. However, when they are the same, they double. More spin means a stronger bond. In fact, the binding energy of diproton and dineutron is negative—it just breaks free! The binding energy of a deuteron is weak, but it does take a high enough temperature to break it.

When it comes to nuclear—or even chemical—reactions, the quantity of interest is the probability of occurrence (or the rate at which the reaction happens). If we know how often a reaction occurs, we can easily make predictions about how much of each element we should find in the universe or after the reaction. The rate depends on three things: the energy needed (the binding energy and masses of the reactants and products), the energy available (the temperature of the surrounding radiation), and of course the availability of the reactants. We know these energies for the nucleons, so we can calculate the reaction rates and estimate how much of each we should expect to find in the universe.

Protons can turn into neutrons and vice versa. Due to the slightly larger mass of neutrons, the process is asymmetric. More mass means more energy, so neutrons are less abundant. There are more protons than neutrons in the universe (we can measure it to be about one neutron for every seven protons), and this ratio was established in the first second of the Big Bang. Why did neutrons stop decaying into protons if those are more favorable? Well, they would have, but they were now bound in the nuclei of atoms. If it were not for the formation of elements, there'd be no neutrons at all!

When the temperature dropped low enough in the first minute after the Big Bang, deuterium started to form. The temperature was low enough that the deuteron bond could not be broken. Now we can start building the bigger elements. Right away, the most stable element, helium-4, started to be built up, and the race was on. But it was over before it even started. The larger nuclei required more energy, and the temperature was dropping. What's more, the number of neutrons available for further reactions was too low. In fact, after only a few minutes, all the neutrons created in the Big Bang ended up as helium-4 (with a few in the next heaviest element, lithium). The extra protons that were left over? Well, they were just hydrogen nuclei, including those in each water molecule in your body.

The concept of the forging of chemical elements in the hot, fiery first minutes of the universe is seen as a great success of modern cosmology. The details were pieced together in the 1940s and 1950s. This was a time when there was an explosion in the study of nuclear reactions, driven by the development of nuclear power and, unfortunately, nuclear weapons. In laboratories around the world, nuclear scientists were working on measuring the various rates of nuclear reactions and understanding the conditions needed to make the nuclei of atoms break apart or fuse together. With paper and pencils and the first true electronic computers, others were working to solve the fiendishly complicated equations of quantum mechanics that govern how particles and atomic nuclei interact. Cosmologists could raid this treasure trove of nuclear data and apply it to their questions about the universe.

A Cosmologist's Playground

The cosmologists' calculations were composed of two main pieces. The first is related to the expansion of the universe. From Einstein's general theory of relativity, we know that expansion depends upon the amount of matter, energy, and radiation in the universe.

The second involved nuclear reactions. These depend upon the temperature and density of material, something we can learn from the cosmological equations. Once we have established the variables, the calculations of the forging of elements are relatively straightforward, simply taking the amount of one element and working out how much is created and how much is changed into other elements at each instant of time.

The mathematics you need to study the nucleosynthesis of elements in the early universe, known formally as a set of *coupled differential equations*, are found throughout science, engineering, economics, and, in fact, just about any field where you want to study change over time. They have even been used to study humanity's response to a (hypothetical) zombie apocalypse (which is an excellent training example for studying disease outbreaks).[4]

Computers are excellent at helping scientists solve such rich and complex systems of equations, and calculations that were done laboriously by hand in the 1940s are now completed in mere instants on a modern computer. With a little computer programming, anyone can recreate the forging of the heavy elements. Even more exciting, you can play with the universe and modify the underlying features, such as the expansion of the universe or the makeup of matter. If you are brave, you can even play with the laws of quantum mechanics and adjust the ways that particles interact. Try it. It really is quite fun! (Although a physicist's definition of fun might not be the same as everyone else's!)

But no matter how much you play with the properties of the early universe, one thing rapidly becomes apparent. In the hot, dense environment of the Big Bang, the forging of elements is very inefficient. The deuterium bottleneck really does put a halt to the formation of the elements, and nucleosynthesis leaves the universe as mainly hydrogen, a smattering of helium, and a trace of the other elements. You have to *really* mess with the makeup of the universe to radically change this outcome.

Cosmologists eventually came to realize that the Big Bang as we understand it could not account for all the different elements that we see around us today. The rapid cooling of the universe as it expanded and the delaying effect of the deuterium bottleneck mean that after forging helium and a little bit of lithium, the universe should have run out of steam. Cosmologists were left wondering where the other elements—like carbon, oxygen, gold, and uranium—came from. There was an obvious place to look that possessed the extreme temperatures and densities necessary to forge new elements—the hearts of stars! However, the physics of these environments was as strange and exotic as those in the Big Bang. More work with more equations was necessary, and smart minds were needed to work out just what had happened.

Following the first few minutes after the Big Bang, the cosmological nuclear furnace dimmed as the universe continued to expand and cool. Leftover radiation also cooled, with the universe eventually fading into blackness. In the dark, gravity dominated, pulling matter together into lumps and clumps. Mass, in the form of the dark matter, the dominant mass in the universe that lurked in the background of the Big Bang, formed the seeds of the first galaxies. Normal matter (the atoms from creation) came along for the ride. The gas cooled and collapsed, crushing down hard and driving temperatures at their cores to extreme values. The first stars were born, and the universe lit up and entered its modern age. At center stage, the world of the quantum was found to play a leading role, for without it, stars would not shine.

PART 2

The QUANTUM of COSMOS PRESENT

How did we unravel the chemistry of the heavens?

We're going to start this chapter with a little history.[1] In 1835, French philosopher Auguste Comte was pondering the nature of the universe. His conclusion was that the makeup of the heavens would remain forever a mystery. In his *Cours de la philosophie positive*, he wrote, "On the subject of stars...we shall never be able by any means to study their chemical composition or their mineralogical structure."[2]

In science, making predictions about the future is a dangerous business. History is littered with now seemingly laughable visions of the future. As we will see, Comte was also wrong about our abilities to divine the constitution of stars.

Almost two hundred years before Comte put pen to paper, the great scientist Isaac Newton took the first steps to unravel the nature of the heavens. In the 1660s, in his rooms at Trinity College in Cambridge, he directed a narrow beam of sunlight into a glass prism. Astoundingly, the white light of the Sun was dispersed into the colors of the rainbow! We can see these little rainbows all around us, if we look hard enough, from sunlight passing though glass that can act like a prism.

By the early 1800s, Bavarian Joseph von Fraunhofer had refined the art of making high-quality prisms and combining them with telescopes. By dispersing the light from bright stars, he found similar rainbow patterns to that of the light from the Sun. Perhaps the Sun and the stars were not so different!

Compared to today, rules on health and safety in the nineteenth century were rather lax, and working with poisonous metal vapors probably contributed to Fraunhofer's death at the age of thirty-nine. While his life was short, with his precision optics, he provided the next leap in understanding the constitution of stars. Examining the rainbow of the Sun's light, its *spectrum*, he found it contained numerous thin dark bands, blacking out very specific colors. While these dark bands in the spectrum of light from the Sun had been noted by William Hyde Wollaston a decade earlier, Fraunhofer began to systematically map out the lines, identifying almost six hundred individual dark bands.

The source of these lines remained a mystery until the 1850s with the work of Gustav Kirchhoff and Robert Bunsen. They directed beams of light though samples of gas, then dispersed them with a prism. They found that the passage of light through the gas had imprinted a series of dark lines onto the spectrum. Different samples of gas produced their own unique fingerprints of bands.

It became clear that the dark lines on the spectrum of the Sun were due to the various elements in its atmosphere, elements that were also found in laboratories on Earth. By the 1860s, the spectroscopy of other stars undertaken by astronomical pioneers like William and Margaret Huggins revealed that more distant stars were also made of earthly stuff.

Well, almost. As well as observing the outer layers of the Sun, some astronomers were taking spectroscopic observations of its tenuous outer atmosphere. This was only possible when the Sun's glare was blocked during a total eclipse. Here, dark bands were replaced by bright lines, explained by Kirchhoff and Bunsen as emissions of light from the elements. Surprisingly, the light from the Sun's atmosphere displayed a bright yellow line that appeared to have no counterpart in laboratories. Perhaps the heavens were, at least in part, composed of non-earthly stuff!

In the 1860s, astronomer Norman Lockyer and chemist Edward Frankland decided that the presence of this bright line indicated the existence of material yet to be identified on Earth. They concluded that something was missing from the periodic table of elements. They named it helium, after Helios, the Greek god of the Sun, and by 1900, scientists had finally isolated this element in the laboratory. By 1903, they were extracting helium from under the ground, finding the first reserves trapped in rock under a field in Dexter, Kansas. They knew what it was precisely by its quantum fingerprint, exactly that found in the lab and in the Sun. Today, we are suffering a shortage of helium as this stable, light gas has found uses in magnetic resonance imaging (MRI), rocket engines, and far too many party balloons![3]

So Comte's prophetic words were shown to be wrong. Spectroscopy revealed the chemical composition of stars and showed that the heavens were built from nothing but earthly elements. With that fact established, the universe became a lot less mysterious.

While the presence of the elemental fingerprints in starlight opened up the details of the universe, why elements possessed such a fingerprint in the first place remained unknown. Why should one element's pattern of bands be distinctly different from another? Through the 1800s and into the early 1900s, chemists and physicists were starting to pry apart atoms, exposing their inner secrets. It is through their story that we will understand how astronomers were able to reveal the elemental makeup of the heavens.

A Quantum Rainbow

When looking at a rainbow, you may notice that some colors appear brighter than others. This is partially because the human eye is not a perfect detector, and it is more sensitive to some colors than to others. But more than that, light from any source is not likely to contain the same amount of intensity for each color contained in it. Spectroscopy is more than just using a prism to look at the colors contained in light—it is about the intensity of each of those colors. So while the sun appears white to the human eye, it actually contains many colors, each with a different intensity. (Don't look at the sun to try and test this, by the way. Every book that discusses astronomy and the Sun has this warning. Don't say you haven't been told!) This is exactly the result a spectrometer produces. If you google "spectrum," you will see a host of images of the colors of the rainbow. However, google "spectrum from a spectrometer," and you will get the laboratory view. Sure, it's not as pretty, but it does reveal hidden information and a beautiful puzzle.

While the dark lines in the spectrum of sunlight did indeed puzzle scientists of the nineteenth century, the rest of the spectrum was also still unexplained. Why are there certain intensities for certain colors? In fact, the puzzle was far more intriguing than that. It was more than just sunlight that appeared to have a preferred spectrum. Every hot object, from glowing iron to burning wood, had a spectrum that seemed to depend only on the temperature it was heated to. It didn't matter what it was—if it was heated to the same temperature, it would glow with exactly the same colors at the same intensities. How could these lines in the spectrum possibly be explained if the spectrum itself couldn't be?

The search to explain the spectrum of light from hot objects had coalesced around a specific theoretical model. It was theorized that the object produced light from a huge collection of oscillating charges. Why that hypothesis? Because it was already known that light was an oscillating electromagnetic wave, produced by oscillating charges. Indeed, this was exactly the principle behind the extremely successful applications of Maxwell's electromagnetism. The speed of wiggling of the oscillator was the energy it had and also the color of light it produced. So the task was relatively simple: find a principle by which the oscillators wiggled in the right proportions to produce the observed spectrum of light. The answer finally came in 1900, with Max Planck, whom we met in the introduction to this book, and quantum physics was born.

Though quantum theory and cosmology are intimately linked, the histories of their development do not line up with the chronology of the universe they have revealed. So far, we have been traveling chronologically in time from the beginning of the universe to today. But we have jumped all over the history of scientific discovery. We have met some of the quantum cast—Einstein, Heisenberg, Pauli, Noether—but it's time to reacquaint ourselves with the father of quantum physics, Max Planck. His Nobel Prize in Physics reads

"In recognition of the services he rendered to the advancement of Physics by his discovery of energy quanta."[4]

It is time to look a little more deeply at how Max Planck kicked off the quantum revolution. In 1900, Planck put forth the quantum hypothesis, that energy came in discrete chunks—the quanta—rather than as a continuous wave.[5] While other physicists tried desperately to create a mechanism producing the characteristic spectrum of light, Planck started playing with what he called mathematical tricks. One trick was to assume that the energy of each oscillator could not take on any arbitrary value but must come in discrete units. There was, in fact, a smallest unit of energy, and that was the quantum. Planck didn't like this idea because it clashed with the classical physics of his education, but it worked. Soon the idea spread to other unexplained phenomena, and quantum physics was out of the gates.

Around the same time, the structure of the atom was starting to be revealed. Scientists only knew at the time that atoms had a dense nucleus of positive charge and that electrons were more diffusely spread around the outside. A popular model depicting the atom was the so-called planetary model, wherein electrons orbited the nucleus just as the planets orbit the Sun. While this thinking is flawed, it is still a useful picture to have in mind even today. After all, we still call the states of electrons "orbitals." The major flaw in thinking of electrons orbiting like planets around the Sun was an obvious one: moving electrons radiate energy. If energy is lost, the electron should also lose speed and spiral, almost instantaneously, into the nucleus, and the atom will then cease to be. So the best model of the atom showed that matter was so completely unstable that none of it should exist at all! Clearly new ideas about the atom were needed.

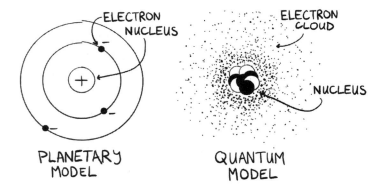

PLANETARY MODEL

QUANTUM MODEL

Jumping to Conclusions

At the time, the global center of research into quantum physics was in Copenhagen—in particular, at the house of Danish physicist Niels Bohr. Using Planck's quantum hypothesis as inspiration, Bohr suggested that the electron should not be able to occupy just any orbit around the nucleus but only certain fixed orbits. Since it could not visit the space between orbits, the electron could not radiate and lose energy. Matter was stabilized—at least theoretically.

While the electron could not visit the space between orbits, it could change orbits, and it did so by *jumping* between them. The electron—says the Bohr model—disappears from one orbit and reappears in another instantly. However, there is still a difference in energy between various orbits. Where did that energy go to or come from? Light. When an electron jumps from a higher energy level to a lower one, a single quantum of energy is released as light. The energy is related to the oscillations of the electric field as Einstein predicted when he applied the quantum hypothesis to light. That is, the energy of the light is directly proportional to the color. Since there are only certain levels of energy allowed and all atoms of the same species are identical, the light coming from them is always the same discrete set of colors.

For example, if you energetically excite a cloud of helium, soon after, it will start to emit light, but it will only emit specific colors—the lines in the spectrum. These energies correspond exactly to the differences between the energy levels allowed by Bohr's model. The allowed energy levels are different for each element since they have different nuclei. Therefore, the fingerprints of elements are inked by quantum physics.

Going in the other direction, the electron can *absorb* light. However, now it is a bit trickier. For the electron to move from one level of energy to a higher one, it must jump there by absorbing exactly the correct amount of energy. Reversing the lines seen in the emission spectrum—that is, sending light back with exactly the same color—allows the electrons to move to higher energy orbits. Or we could shine light of all colors on the atoms. What happens then is that only those colors that correspond to allowed energy transitions are absorbed. The rest pass through unnoticed. At the other side, we see the same light we started with, except missing lines in the spectrum corresponding to those special colors. These are the dark lines in the absorption spectrum.

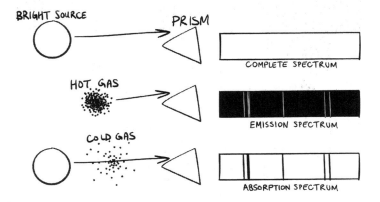

We've covered a lot of ground here, so let's recap. Orbits of electrons in atoms are quantized, meaning they can only have particular energies—they are discrete rather than continuous. For an electron to jump from a lower energy orbit to a higher energy orbit, the atom absorbs a photon with the right amount of energy to make the jump. This results in an absorption line at a specific frequency in the spectrum of light. When an electron falls from a higher energy orbit to a lower energy orbit, the atom emits a photon of light with a specific frequency. This results in an emission line in the spectrum of light.

Looking at the Sun (Don't Try This at Home)

The hot interior of the Sun produces light of all colors, which is why it appears white to our eyes. As that light passes through the less dense outer layers of the Sun, the quantum jumps start to happen. Each photon that has the right energy for a jump gets absorbed. It also quickly gets emitted, but now in a random direction. The net effect for the telescopes on Earth is a spectrum with absorption lines corresponding exactly to the energy of the jumping electrons. This also explains the emission spectrum that is seen by a telescope not pointed directly at a bright light source. Those emitted photons can be seen against an otherwise dark background. Quantum jumps also allow us to see the fingerprints of the atoms in the outer atmosphere of the Sun during an eclipse and those of interstellar dust clouds.

The Bohr model for the atom required many refinements before ultimately giving way to the full quantum treatment, with wave functions and probabilities, something only possible once all the mathematical elements of quantum mechanics were finally in place. However, the one feature that remained was the discreteness of atomic energy and its interaction with light. Quantum physics explained the spectral observations of astronomers. But more than that, each new theoretical method in the quantum theory of light and matter suggested a new way to interpret the data from the stars. This ushered in the new science of *astrophysics*, the scientific discipline that examines the life cycles of stars, planets, and other objects in the universe. Scientists were no longer limited to mapping out the positions of the planets and stars but could now start to understand their very nature.

The discovery of the quantized energy levels of atoms gave birth to the modern era of quantum mechanics, and the understanding of atomic structure has revolutionized astronomy and cosmology. Every night, around the world, telescopes are trained on the heavens. Telescopes that aren't limited to the optical part of the electromagnetic spectrum, such as those using radio or millimeter wavelengths, can carry on observing in the daytime.

Telescopes are really undertaking two main tasks. The first is imaging—literally taking a picture of the sky. We can learn a lot from these pictures, such as how many stars are in the galaxies and how many galaxies are in the universe. But if we look through filtered glass, where we can compare what we see in blue light to that seen in green or red, more secrets are revealed. From the color of a star, we can deduce its temperature, and from the color of galaxies, we can determine the life cycles of stars living within.

The real power of telescopes, however, is spectroscopy. Recreating the experiment of Newton and his prism but on a much larger scale, astronomers disperse—or spread—the light from distant stars and galaxies. Glass prisms are rarely used though. Modern astrophysics relies on devices called *dispersion gratings* to more efficiently achieve the same results. A great example of a dispersion grating is a compact disc. It can be incredibly difficult to see the spectrum even with a good prism, because the light has to be at a very specific angle and change media twice (air → glass, glass → air). But take a quick glance at a compact disc in almost any lighting conditions and you will see a brilliant display of colors.

What do astronomers look for in the dispersed light of distant objects? Well, this rainbow is rich in information about the source of the light, showing which glow simply due to their unique temperature (like stars) or display more complex emissions from superfast and superheated material (like the matter swirling around supermassive black holes in active galaxies known as *quasars*).

Overlapping the rainbowlike emission from a source is the bar code of lines due to electron transitions in atoms. In stars, these are generally seen as lines of absorption, where atoms in the atmospheres of stars absorb well-defined frequencies of light due to their electron transitions. Sometimes, dependent upon the physical state of the atmosphere, electrons falling from higher energy levels emit photons of light, producing a line of emission rather than absorption.

Quasars are some of the brightest objects we know of, observed right across the universe. With telescopes and spectroscopy, astronomers have been able to unravel the nature of these luminous beasts. At their hearts sit black holes that can be a billion times the mass of the Sun! The black holes at the very center are, well, black, but surrounding them are rapidly rotating disks of matter. Heated through friction, these disks glow brightly, illuminating immense gas clouds that orbit nearby. This heating excites individual atoms, with their electron transitions resulting in bright emission lines, with prominent features from hydrogen and carbon.

The light from these distant quasars has to traverse many billions of light-years through space to get to us. This space is not entirely empty. Scattered among the galaxies are immense clouds of gas, mainly hydrogen, and, like most material in the universe, polluted with the heavier elements formed in stars. As the quasar light travels through the universe, clumps of hydrogen eat into the spectrum of light, leaving a pattern of distinct absorptions, their locations dictated by the ever-expanding universe.

The quantization of electron orbits and the precise bundles of energy that are absorbed and emitted as electrons undergo their transitions gave astronomers a new window on the universe. This provided them with the ability to determine the chemical composition of objects across the universe, a scientific miracle. The mysterious matter of the universe was shown to be nothing more than mere earthly material, from the nearest stars to the edge of the observable universe. And if the stuff up there was just like the stuff down here, we could use our earthly laws of physics to understand how it interacts and changes. From Comte's failed prediction about the unearthly nature of matter in distant reaches of the universe, telescopes, prisms, and oscillating and jumping electrons finally brought the composition of the heavens into our grasp.

Where did the chemicals inside us come from?

In Part 1, we looked at the formation of chemical elements in the fires of the Big Bang. This process was impeded by the fragility of deuterium, a bottleneck that resulted in the universe being too cool when nucleosynthesis began to appreciably form elements heavier than lithium. The cooling early universe was a soup of simple chemical elements, but today there are many more, from barium to uranium. Elements heavier than hydrogen and helium are essential for our existence. But just where did these elements come from?

After the initial cosmic fires were extinguished, the universe descended into an eerie darkness. It was a time before stars. The hot soup of fundamental particles had been replaced by a tepid soup of protons, the nuclei of hydrogen atoms, and the nuclei of the few lightest elements. Accompanying these were free electrons, the temperatures being still too high for them to join with the atomic nuclei. After four hundred thousand years, in an event that astronomers refer to as *recombination*, the nuclei and electrons combined. This is a confusing title, as the nuclei and electrons were never combined previously!

In the darkness, gravity was at work. Remember that the matter in the early universe was not completely smooth but seeded with subtle differences in density from quantum fluctuations left over from inflation. Gravity pulled matter together, pooling into regions of growing density and forming immense clouds. Within these clouds, the density continued to increase as the gas cooled, losing energy as it radiated. These clouds fragmented into massive chunks weighed down by their own gravity, undergoing collapse and forming the first clutch of *protostars*.

Initially, these protostars glowed feebly in the darkness, heated by the compression from the ongoing gravitational collapse. Gravity continued its squeezing, and the protostars collapsed further. The central regions of the protostars were squeezed hard by the weight of their outer regions pressing down. The temperature and density within the protostars' cores began to soar, with collisions starting to bring atoms closer and closer together. Eventually, the densities and pressure reached immense levels, and electrons were torn from their atoms. Within this plasma, atomic nuclei were again forced close together, close enough for the strong force to reach out and bind them together. Nucleosynthesis began again as the stars fused lighter elements into heavier elements. Through this forging of elements, energy was released. This nuclear energy radiated from the core and through the outer layers of the star, providing a force that would counter the gravitational collapse and support the star through its lifetime. Roughly five hundred thousand years after the Big Bang, the nuclear energy burst from the surface of the first stars, illuminating the universe.[1]

Despite many similarities, there is an important difference between the conditions at the hearts of stars compared to the early universe. This difference has a big impact on the forging of elements. The early universe was an almost equal mix of the two nuclear particles, protons and neutrons. The first stage of forming heavier nuclei was the binding together of one of each, a single proton and a single neutron, to form deuterium. Once there were appreciable amounts of deuterium, pairs of the atoms could be forged into helium. But stars lack the free neutrons needed to form deuterium. Remember, any neutrons not locked away in the first few Big Bang elements decayed away rapidly into protons. So the core of one of the first stars would have been mainly free protons with a few other elements thrown into the mix. The physics inside a star is the same as that at the Big Bang, and while protons can get close enough to feel the strong force, we saw that this combination, known as a diproton, is unstable and instantly falls apart.

Without a route to creating deuterium, it would appear that the first steps to powering stars is cut off. So just how do stars overcome this second deuterium bottleneck?

The formation of deuterium is not the only bottleneck in forming elements in stars. Naively, you might think that all you have to do is crash lighter nuclei together to build a bigger nucleus, but of course, it's more complex than that. Some combinations of protons and neutrons, especially those having too few or too many neutrons, are unstable and fall apart in an instant. Also, if the collision is too energetic, a new, heavier nucleus might form, but the internal motions of protons and neutrons sloshing about might be enough to rip the nucleus apart into lighter elements.[2]

Given all this, it might seem that forming elements in stars is a tortuous affair. It appears to require seemingly impossible conditions just to get going and then some "just right" conditions in terms of energy to proceed. So while the Big Bang furnished the universe with the simplest elements, we still have to wonder exactly where all the other chemical elements, the ones that make up you and me, came from.

The Quantum Shortcut

To understand this phenomenon, we'll have to head back into the mountain range of energy. Stuck in a high valley, you seem to need a catalyst to turn potential energy into kinetic energy, allowing you to overcome the next peak. Since the only free nucleons we now have are protons—the neutrons were essentially locked away in helium-4 from the Big Bang—they seem to need a lot of energy too. Why? Remember, protons are positively charged, and if two slowly approach each other, they will be repelled by electrostatic repulsion. They face a mountain of potential energy that needs to be overcome. We already discussed the aptly named strong force, but it is no help here as it only acts on distances the size of particles. At all the other scales, the electromagnetic force reigns supreme.

Consider, for instance, where you are sitting right now. One of us—we won't reveal which—is writing this sentence while seated on an uncomfortable bench. Perhaps you are happily nestled in a comfy chair. In either case, neither we nor you are actually touching anything. That is, *your* atoms are not touching the *chair's* atoms. In fact, at the atomic level, you never touch anything! How can that be? It all comes down to the electric force. The electrons orbiting your nuclei repel the electrons orbiting the chair's nuclei. Put as much weight as you want on that chair—you will never get the atoms to touch. The electrostatic force is *that* strong!

In other parts of the universe, way more force than our measly bodies can provide is at play. Inside stars, atoms have to touch not just by overcoming the electrostatic repulsion of electrons. Once atoms get close enough, their nuclei must also overcome the electrostatic repulsion of the positively charged protons. The mountain of energy they must ascend is a very steep peak. The kinetic energy needed to do it is so large that even if it was overcome, the protons would just bounce off each other on the other side, so to speak. So how do they do it? The answer lies in one of the features of quantum physics that renders almost all classical ways of doing calculations obsolete. It's called *quantum tunneling*.[3]

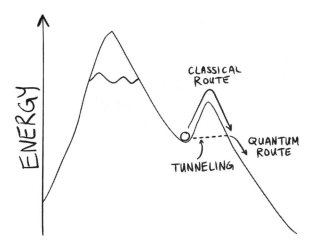

Physicists often lament the counterintuitive nature of quantum theory. However, quantum tunneling is one of the few things in quantum physics that is actually easy to comprehend. Indeed, it is exactly as it sounds. Facing a mountain of energy, instead of going over, you can go straight through—you can tunnel.[4] However, it's not easy to tunnel, nor is it a guaranteed method of success. In most cases, the chance of a tunneling event happening is small, so physicists talk only about tunneling *probabilities* or tunneling *rates*. To get a sense of the scale of importance of tunneling, note that all reaction rates calculated in nuclear physics and chemistry are essential tunneling rates. All processes rely on this shortcut rather than climbing the high peaks of the energy mountain. Of course, for an *individual* atom or molecule, the time it would take to successfully tunnel might be the age of the universe. However, chemistry is about *lots* of identical atoms or molecules all trying to do the same thing. It's like when a lot of people are doing something that succeeds only rarely; chances are at least a few will win. A casino talks about the chances or rate of winning because it cares what happens in the aggregate, not what happens to individual gamblers.

The idea of quantum tunneling provides a curious bridge between classical and quantum physics. While the calculations of classical physics no longer apply—and indeed suggest things like tunneling through an energy barrier are impossible—the *ideas* of classical physics are still useful. Classical intuition and physics can take us a long way toward building a mental model of what's going on: that we are in a valley, facing an insurmountable peak, and at the last minute—when all seems lost—quantum tunneling takes over and reveals the solution. Much of what we think about in quantum physics is grounded in classical language. In the case of tunneling, we can think of quantum physics as adding a little bit more power to what classical physics allows. Unfortunately, the tunneling trick only works on quantum scales, not human scales.

An Impossible Superhero Power

Suppose you have somehow ended up on an obstacle course. You are facing a wall. To win, you must get to the other side of the wall. Whether or not you are thinking about it that way, this is a physics problem. Your body needs to muster enough kinetic energy to match and overcome the potential energy you would have at the top of the wall. *Ah, but wait,* you think. *What about going right through the wall by quantum tunneling?* Indeed, you could do that. There is a chance, by running straight at the wall, you will end up on the other side. But before you try, know that the odds are unfathomably small. You could run at walls your entire life, and even if you lived to see the end of the universe, you probably would not have succeeded in tunneling. It might happen, but this is an extracosmic scale bet. Also, it might hurt a lot.

The chance of an object tunneling depends on a few things: how big the barrier is, how much energy the object has, how far it must go, and how big the object is. The larger the object, the less likely a successful tunneling attempt. By the time the size of the object is big enough to be visible to our eyes, the chance of tunneling through a barrier is essentially zero—close enough to zero to call it impossible. Thus, we never notice tunneling in our everyday lives of big objects—when you sit down on your chair, you know it is going to provide support—but if the rules of quantum mechanics applied on large objects like a human body, there would be a chance that you would occasionally pass right through the chair and end up on the floor or even below it! But for little things like protons and electrons, tunneling is the preferred mode of transportation. If you are going to write a superhero comic about tunneling, make sure your character is microscopic!

The Quantum Goldilocks Zone

We opened this chapter with a discussion about the power of the Sun and its role in the creation of the elements. Remember, we are building the elements from the ground up—one nucleon, then two, then three, and so on. Our first step on the road to creating the heavier elements with many nucleons was creating deuterium, a proton and neutron bound together, the smallest composite nucleus. But in the core of the Sun, we lack any neutrons, and all we have are protons crashing together. Because they are positively charged, they strongly repel one another, never getting close enough for the strong force to grab hold.

Each individual proton collides with other protons many billions of times every second without combining, but in this maelstrom of collisions, every so often, two protons can tunnel through the electrostatic energy barrier that is keeping them apart. Then they suddenly find themselves close enough for the strong force to try and hold them together. They have formed a diproton!

But as we have seen, diprotons are unstable and instantaneously disintegrate back into two protons. It seems that ultimately nothing has changed! But there is one more force at play, the *weak nuclear force*.

The weak nuclear force can play a trick that no other force can: it can change protons into neutrons! But the chances of this are very small. So if we can use tunneling to form a diproton, there is then a small chance that one of the protons will convert into a neutron, forming a stable deuteron before it can fall apart. The chances are very, very small, with only about one in every 10^{28} (10 octillion) collisions between protons in the Sun producing deuterium. It's a highly inefficient process, but it is the first step to creating heavier elements.

Our journey is not over, however, as tunneling is not a silver bullet in forming heavier elements. Yes, it acts as a catalyst to get over the electrostatic repulsion, but other variables are also at play. Think about making lemonade: water plus lemon juice plus sugar. If you put too little sugar in, it will be sour; too much sugar, and it will be too sweet. There is a Goldilocks amount of sugar, but too little or too much still makes lemonade you can sell at the stand. The Goldilocks zone of energy in nuclear reactions is much less forgiving.

Imagine two tennis balls made of hook-and-loop fastener (better known to some by the brand name Velcro). Push them together, and they will stick. Throw the stuck pair at the ground, and all that kinetic energy might be enough to break them apart. Now, separate the tennis balls and throw them at each other. They must be thrown fast enough to reach each other in the first place. At that speed, they will need a head-on collision to stick together—a glancing blow will hardly be noticed by the sticky material. But even if they hit each other head-on, too much speed will cause them to bounce apart. There must be a Goldilocks zone of ball speeds for them to stick. The combined kinetic energy of the balls must not be greater than the energy sufficient to break them apart. Even for this simple task, a successful pairing would be rare.

Atomic nuclei whizzing about in stars behave in a similar way to these funny tennis balls, but what happens is now dominated by chance and ever rarer events. The tennis balls needed a minimum speed to reach each other. The same is true for nuclei. In order for quantum tunneling to succeed with an appreciable probability, they need to have a lot of energy. In stars, this is provided by the pressure caused by gravity, squeezing particles together due to the immense weight of the star. If two nuclei manage to come together, the new combined nucleus will have energy equaling the total energy of the original nuclei. If that energy is way too high, the nucleus will break apart immediately, just as the tennis balls bounced off each other when they struck with too much speed.

The Goldilocks region of tennis ball speeds is probably fairly wide, relatively speaking. For nuclear reactions, however, the target energy is quantum mechanical, and that means there are only specific energies that will work, as Planck taught us way back at the start of the twentieth century.

When the energies of interacting things match, physicists call this a *resonance*. Of course, the concept of resonance is not restricted to nuclear physics. In music, the hollow body of the guitar amplifies the vibrations of the strings. A different size or material would not do this in the same way, but perhaps so subtly that only an expert could detect the difference. The energy of the vibrations caused by the strings matches the energy of the vibrations allowed by the hollow cavity. In this way, your voice is also an example of resonance. Your body pushes air out with many vibrations. Your jaw, lips, tongue, teeth, and other organs change the shape of your vocal tract to amplify specific frequencies. Puckering up your lips doesn't create a whistle—it only amplifies the inaudible whistle from just blowing air. Everyone can whistle, but only some people can amplify it to make the sound we recognize as whistling.

Resonance amplifies particular interactions. Exactly which interactions is a complicated issue, depending on many factors and involving complex, sometimes laborious, calculations. More often than not, the properties of resonances are too difficult to determine from the mathematics of nuclear and particle physics, and the best we can do is just measure them in the laboratory. Today, we can map out the shape of a guitar body and use computers to simulate how it will vibrate, hence determining its resonances. However, with all the complex ways your body can shape its vocal tract, even computers cannot determine what resonances you as a human can create. So it was quite impressive when physicists in the mid-twentieth century—without computers or even a full-fledged standard model—were able to predict resonances occurring in the Sun.

The Tune of the Sun

The hero in this story is Sir Fred Hoyle, one of the most influential astrophysicists of the last century.[5] As we discussed previously, he gave the Big Bang its name (and did not mean it to be a compliment) and is well known for his role in popularizing science and writing science fiction. He is also known for some of his more "out there" scientific ideas on the origin of life and the nature of the universe. But he is most famous for his understanding of how stars work.

In the early days of nuclear physics, resonances were hypothesized by necessity from the simple fact that we exist—the first instance of a so-called *anthropic argument*. For example, we know that carbon exists because humans—and many other things in the universe—are made of carbon. Therefore, there must be a pathway for carbon to be created in stars. From our understanding of the properties of atomic nuclei, we can calculate these pathways to heavier elements in stars and calculate the expected abundance of elements in the universe. When scientists first attempted to define this theory in the early 1950s, it was apparent that to account for the universal abundance of carbon, there must be a resonance at an appropriate energy that boosts its production.

Indeed, in 1954, through this line of reasoning, Hoyle predicted a new energy level of carbon by arguing that such a resonance must be present for three helium nuclei to eventually create a stable carbon atom. Experimenters had already found many resonances of the carbon nucleus, but a resonance at the particular energy predicted by Hoyle appeared to be absent. Hoyle was not one to give up, badgering experimenters to look harder. Soon enough, they confirmed Hoyle's prediction.

So atomic nuclei, like our tennis balls discussed earlier, have a very narrow window of opportunity to stick, and the Sun is inefficient in turning lighter elements into heavier elements. In some sense, this is good. This nuclear burning of light elements produces sunlight, the energy that fuels life on Earth. But the difficulty—or low probability—of the reactions that create heavier elements allows us to live comfortably on Earth. If it were too easy for these reactions to occur, the Sun would burn up its hydrogen fuel much more quickly, and we would not have the stable energy it has provided our planet for hundreds of millions of years.

It is sobering to think that the carbon in our bodies and the oxygen we breathe were formed in the hearts of previous generations of stars, stars that lived their lives over billions of years before our Sun was born. Heavier elements, such as the gold in our jewelry, were created in some of the most extreme and violent events in the universe, at the ends of the lives of stars, a point we will return to shortly. But the process is the same: get nuclei that repel each other close enough that the chance of quantum tunneling gets them over the last hill, and let the strong force bind them together.

From the atoms that define the material world around you, including your very being, to the sunlight that warms your skin on a summer day, all this is possible because of the quantum.

Why do dying stars rip themselves apart?

Stars burn by forging heavier elements from lighter elements. The rate at which these nuclear fires burn depends upon the conditions in the heart of a star. Simply put, the higher the density and temperature, the more rapidly elements are transmuted and the more brightly a star can shine. For an individual star, these characteristics are defined by its mass.[1] The more massive a star, the more gravity can squeeze the core to higher densities and temperatures and the more energetic the stellar output.

In the smallest stars—the ones that barely achieve the conditions to ignite their nuclear reactions—hydrogen is converted to helium in a very sedate fashion. With a mass only about one tenth that of our Sun, these *red dwarfs* glow feebly but have a hundred trillion years of fuel to burn through. Once the hydrogen fuel is gone, the core of the red dwarf is too cool to burn helium into heavier elements, and the star simply blinks out, cooling and fading into the darkness.

Our Sun, being more massive, can squeeze its core harder. It could burn through its nuclear fuel in a mere ten billion years, but once the hydrogen is exhausted, a little extra squeeze can begin to burn helium into carbon and oxygen. This internal rearrangement will have a profound effect on our Sun, causing its outer layers to swell and cool. During this *red giant* phase of its life, the outer layers of the Sun will swell to engulf the orbits of Mercury and Venus and possibly outward to swallow the Earth and Mars. But don't worry—we have another few billion years before this radical change begins.

Eventually, our Sun and other stars with a similar mass will exhaust their nuclear fuel. The core will become too cool, unable to burn carbon and oxygen into anything heavier. The star will undergo more internal upheaval as the fuel is depleted, pulsating as the nuclear burning becomes erratic. In the end, the outer layers of the star will be puffed off in one final sigh. While the result can be spectacularly beautiful, viewed through telescopes as planetary *nebulae*, they are the markers of stellar grave sites.

The life of a more massive star, several times larger than the Sun, can be yet more spectacular. The immense gravity of these large stars means the conditions at their hearts are no barrier to nuclear burning. Hydrogen is rapidly burned into helium, helium into carbon and oxygen, and then up into heavier and heavier elements. Very massive stars can churn through their nuclear fuel in a few tens of millions of years, constantly readjusting their internal structure as material created in one nuclear reaction becomes fuel for the next.

A star about ten times the mass of the Sun will spend roughly ten million years burning through the hydrogen at its core, then about one million years burning helium. Burning through carbon might only last a few hundred years, while oxygen burning might be over in a few hundred days. The final stage, burning of silicon, takes a matter of hours. Then the nuclear burning comes to a grinding halt.

The result of burning silicon is the production of iron, and iron has a special atomic nucleus. In iron, the protons and neutrons are tightly bound together. If you want to transmute iron into other elements, you need to put significant energy in to overcome this tight binding. This means that unlike other nuclear reactions that liberate energy and allow the star to shine, nuclear reactions with iron suck energy in. Once the star has a core of iron, the nuclear fires are completely extinguished.

Without the radiation pressure pushing outward from the stellar core, there is nothing to halt gravity. The outer layers of the star free-fall inward, crushing the now-dead star heart. As they do, the crushing forces drive the temperature and density into extremes, and there is now enough energy to turn iron into heavier elements. The core of the star is crushed. In the most massive stars, this crushing is potentially into oblivion, creating a black hole, and the outer layers are driven off in a violent explosion. For slightly less massive stars, the result is an immensely dense dead stellar heart, known as a *neutron star*.

During the immense squeezing of the stellar core due to the collapsing outer layers, strange things start to happen. Protons and neutrons get packed together at such high densities that the strong force, which normally holds nuclei together, becomes repulsive, and the infalling outer layers are pushed outward as the star starts to explode. In this superdense, superhot environment, there is so much energy swishing around that even iron can be forged into heavier elements.

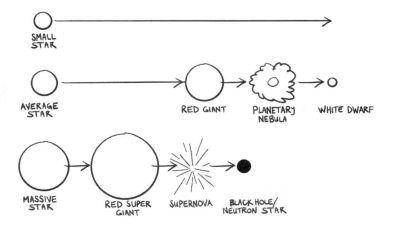

In this case, we have one of the most spectacular events in the universe, a *supernova*, where light from one dying star can, for a few weeks, outshine the combined brightness of the billions of other stars in its galaxy. Supernovae are violent events, but this spectacular end to the life of the star is not dictated by the superheavy elements created in the violence or the intense burst of high-energy radiation. Instead, it is caused by a tiny, strange, ghostly particle that is barely even there, the *neutrino*. How can this little piece of nothingness be responsible for ripping a star apart?

A Recipe for Star Stuff

To understand this process, let's bake a theoretical cake. Mix together ½ cup butter, ¾ cup sugar, 2 eggs, 2 cups self-rising flour, and ⅔ cup of milk. Pour the mixture into a deep cake pan. Weigh the pan of mixed ingredients—let's round it up to 1 kg. Then, bake it for 45 minutes in an oven preheated to 180°C/350°F. After letting it cool, weigh the pan of cooked ingredients—850 g. While you enjoy your butter cake, let's think about why the cooked cake weighed less than the raw ingredients. To do that, let's ignore the deliciously fascinating chemistry of baking and just do the math. If the ingredients and pan weigh 1 kg (1000 g), and the cake and pan weigh 850 g, then obviously 150 g of some of the ingredients are missing. But which ones? Time for an investigation.

The ingredients contain fluids, but the cake is dry. (Not too dry, of course!) And even though water was not explicitly an ingredient in the recipe, butter, eggs, and milk all contain mostly water.

Water evaporates in hot, dry conditions like the inside of an oven. So our hypothesis is that the lost weight is water. In fact, if you were to capture the ventilated air from the oven and cool it, the water vapor would condense back into liquid water, and you would get the 150 g of missing water. Mystery solved!

But wait. What does this have to do with supernovae or physics at all? We have already mentioned the important concept underlying this idea: conservation. In the case of baking, it's conservation of mass that we are interested in. That is, for most everyday situations, mass is never created or destroyed. By appealing to this conservation law, simple arithmetic can tell you where the missing ingredients are.

In the early 1930s, ingredients were missing in the physicist's version of baking: nuclear reactions. Recall that way back at the early stages of this universe (and this book!), a free neutron could decay into a proton. However, a free proton cannot decay into a neutron, which is why there are more protons than neutrons even today. Something doesn't add up though. A neutron has no charge, while a proton has a positive charge. This neutron-to-proton transition must violate conservation of charge. To match the neutral charge of the neutron, the proton must be accompanied by an electron, and indeed it is.

This was never *really* an issue, though, since the additional electron was the first noticeable thing in such a reaction. Chronologically, the story is reversed.[2] The electron was discovered first in the context of radioactivity. This type of radioactivity is where the proton remains in the nucleus of the atom and the electron is ejected, a phenomenon detectable via many experiments. In fact, plenty of properties of the electron can be measured, and it was immediately apparent that even more things than charge didn't add up. For one, the mass and energy of the original neutron was more than that of the produced proton and electron. Like water evaporating in the oven, something was missing.

It was physicist Wolfgang Pauli who first proposed that the missing energy could have been ejected as another particle. Since charge was already conserved, this new particle must carry no charge and be neutral. It was also posited to have very little mass or perhaps no mass at all, much like the photons that make up light. Another physicist, Enrico Fermi, thus named it the "little neutral one" or, in his native Italian, *neutrino*.

This early success of applying conversation laws in predicting the hypothetical neutrino was long before it was eventually detected, in an experiment in 1953, and many years before our now beloved standard model of particle physics was finalized. At the Los Alamos National Laboratory in New Mexico, Frederick Reines and Clyde Cowan created a detector using 300 liters of water (they used water because it is dense, abundant, and nontoxic).[3] When a neutrino hits a water molecule, a burst of gamma rays can be detected. In fact, they also detected *antineutrinos*, and a slew of many other types of neutrinos, called *flavors*, were found later. The standard model contains three flavors of neutrinos and, of course, their antineutrino counterparts.

This is a reminder that the standard model is one of the great successes of science. At first glance, with all its funny jargon, with fermions and bosons, quarks and electrons, it may seem complicated. But it is an incredibly concise summary of (almost) everything we know about physics. It has yet to find a replacement that can do better at predicting the zoo of particles and forces we find at the fundamental levels of the universe, so it remains the best theory we have, though we know it has many holes, which we will come to later. For now, know that there are twelve fundamental particles that make up matter, and three of them are neutrinos. Each particle species has its own unique characteristics, but neutrinos are the only ones that interact via the weak nuclear force and gravity alone.

FORCES OF NATURE
POWER RANKING

		RANGE	RELATIVE STRENGTH
GRAVITY	◯→←○	∞	10^{-36}
WEAK FORCE		10^{-18} m	10^{-7}
ELECTROMAGNETISM		∞	1
STRONG FORCE		10^{-15} m	10^{2}

Gravity is already the weakest force of the four fundamental forces, and the neutrino mass, as far as we can tell, is incredibly small, making it less susceptible to gravity's influence, so we can ignore its effect on neutrinos here. Like the strong force between protons and neutrons, the weak nuclear force in turn has a very short range. Putting this all together means that a neutrino typically travels immense distances before the happenstance event that it smashes into another particle. For this reason, it is colloquially known as the *ghost particle*.

While this might seem frustrating for the scientist striving to detect neutrinos, it is actually quite reassuring, because about one hundred trillion of these little particles are passing through your body every second. In the same second, about a hundred high-energy particles from space, known as cosmic rays, crash through our bodies. These have the potential to significantly damage your DNA, a potential source of cancer. Luckily, given their minuscule chance for interactions, neutrinos pass harmlessly through.

Cooking Up Neutrinos

Now, where do all these neutrinos come from? Everywhere particles fuse or decay, neutrinos can be created. Some of them may have existed since the beginning of universe, when the first subatomic reactions began to occur. Billions come from the Sun, where deep in its core, hydrogen is fusing into helium; one of the important by-products, due to the action of the weak force, is neutrinos. In addition to neutrinos and the more familiar photons, the Sun also sends high-energy protons our way. These and other sources of cosmic rays crash into atmospheric molecules and explode in the same kind of reactions that humans engineer in giant particle accelerators. These reactions cascade in a shower of yet more high-energy neutrinos. Like a scene from a sci-fi movie, we are being bathed in countless particles that simply pass, ghostlike, right through us.

Among this background of neutrinos that are constantly coursing through the Earth are short spikes in neutrino count. These are the signatures of exploded stars. In fact, we might say this quite literally, as neutrinos can arrive long before the observable photons that can occasionally be seen starkly even by the naked eye.

How do exploding stars produce neutrinos? Let's consider a very massive star, something more than ten times as massive as the Sun. We've already seen that the nuclear furnaces in their core can burn from hydrogen to iron for tens of millions of years before the fires go out and the outer parts of the star collapse in. We've mentioned that in these conditions, even heavier elements are forged, but something else is also going on in the crushed stellar core.

Remember, up until the fires go out, the core is an immense ball of iron nuclei. Each iron atom on Earth is accompanied by twenty-six electrons orbiting the nucleus, but in the immense temperature just before the fires go out in a star, no iron nucleus can hold on to its electrons. But these tiny, negatively charged electrons are still there, buzzing around the mix. And once the fires are extinguished, they play a very important role in the destruction of the star.

Once the outer layers crash down on the core, the iron nuclei are forced together, so close that they lose their individual identity. The core of the star essentially becomes a giant atomic nucleus, an immense ball of protons and neutrons. But unlike a normal atomic nucleus, electrons are still present within this mix.

In this crazy environment, in conditions we can never recreate here on Earth, the electrons are forced into protons, combining via the weak force to create neutrons. In each of these little weak force interactions, a neutrino is created as a by-product. A huge number of neutrons, almost 10^{60}, are created, resulting in an enormous flow of neutrinos from the core. And the astounding thing is that physicists have detected neutrinos created in these cataclysmic events.

SN1987A.[4] That probably means nothing to 99.99 percent of living people. But to astronomers, those characters are a familiar sight. (No, it's not Geraint's password—well, not anymore!) This was the name given to a supernova event in the Large Magellanic Cloud, a satellite galaxy to our Milky Way. The name SN1987A gives it away—in late February 1987, for the first time on Earth, neutrinos were detected from a supernova. In fact, it was the brightest supernova seen from Earth in nearly four centuries, a temporary star that could be seen by the naked eye and is still the object of study for many astronomers.

But wait a minute. How did we detect the neutrinos from SN1987A? That is, how did we detect the seemingly undetectable?

First, a quick reminder about how difficult this task might be: a single neutrino might be able to pass through several light-years' worth of solid lead before interacting with a single lead atom. So how the heck are we supposed to catch neutrinos? The answer: with some pretty intense physics experiments. In addition to detectors placed near artificial neutrino sources, such as particle accelerators and nuclear reactors, there are many neutrino observatories looking for cosmic sources of high-energy neutrinos. One such example is the so-called Super-K, or Super-Kamiokande, in Japan, a neutrino detector submerged in over fifty thousand tons of pure water buried a kilometer under the ground. Another is the aptly named IceCube at the South Pole, buried deep in the Antarctic ice. These and other neutrino experiments go to the extremes of science and engineering to find neutrinos.

Any signal scientists hope to find when trying to detect cosmic neutrinos will be tiny. Since neutrinos mostly pass through the Earth, the ground above deeply buried detectors acts as a convenient shield against all the other particles that would otherwise drown out the neutrino signal. But even deep underground, the neutrino signal is small. In the case of SN1987A, three neutrino observatories detected a whopping twenty-five neutrinos. Now, of course, twenty-five neutrinos sounds like a pittance next to the one hundred trillion we know pass through your body each second.

But the energy of these extra twenty-five neutrinos and the fact that their arrival coincided with one other and the observation—via conventional astronomy—of SN1987A provided convincing evidence that the source of these neutrinos was the core of the collapsing star.

Neutrinos are on double duty in the heart of a dying star. As we mentioned, they arrived before we saw the SN1987A event with our telescopes—three hours before, in fact! And indeed, precisely because of this, the SuperNova Early Warning System is a since-created network of neutrino observatories designed to detect the earliest signals of nearby supernovae. Neutrinos arrive before light and matter precisely because many are able to pass through the dense iron core of the dying star completely untouched. Light and matter interact more strongly and are impeded by the core, taking far longer to escape into interstellar space. However, it is the sheer number of neutrinos that ultimately causes the explosive outflowing shockwave, as only a small fraction of them are necessary to heat the ejected material and interstellar gas.

When a star explodes, a huge number of neutrinos are produced in the runaway nuclear event of a supernova. But the chances of any individual neutrino interacting with the stellar atoms is small, and most directly escape and zip off into the universe. But there are an unfathomable number of neutrinos, so even a tiny fraction crashing into atoms carry an almighty punch—a punch big enough to tear an entire star apart.

Once the spectacular show is over, there might be little remaining of the star that went supernova. As we've mentioned, in the most massive stars, the core is eventually crushed out of existence, forming a black hole. We will return to these exotic objects later.

For less massive stars, the core remains. It has been immensely crushed, with more than the mass of our Sun squeezed into a ball only twenty kilometers across. These objects are made almost entirely of neutrons packed together. Unimaginatively named *neutron stars*, these are some of the most extreme objects in the universe, with an absolutely crushing gravity at their surface, a hundred billion times as intense as gravity on Earth. We really don't understand the details of neutron stars, and their inner cores might be so extreme that even neutrons are ripped apart and free quarks roam. But we can see them as *pulsars*, blinking at us with regular bursts of radio waves, scattered throughout the galaxy. They will eventually cool and fade into the darkness on timescales much longer than the current age of the universe.

But a mystery remains. Inside a neutron star, there is no nuclear burning. Unlike a normal star, it is not being held up by the flow of energy from the nuclear core. So what is there to stop gravity from winning the battle and crushing the neutron star out of existence? As you might have guessed, the quantum will come into play here, but that is a story for later in this book.

Is the entire universe a quantum thing?

How do we describe the entire universe? At first glance, this might seem like a very strange question, but to understand why this concept is important, we have to think like a physicist.

What physics is and what physicists do can be a little hard to define. But it is useful to think of them observing and experimenting on the natural universe and explaining what they see in terms of rules and laws. In textbook terms, it is the observers and the experimentalists who probe the natural world with telescopes, microscopes, and oscilloscopes. Uncovering these laws is the role of the theoretician, someone skilled in the language of mathematics and how to manipulate equations to describe the physical world around us. However, this clean divide does not necessarily mirror reality, and many scientists have feet firmly in both camps.

Isaac Newton, one of the greatest modern scientists, was skilled in both experiment and theory as well as writing on alchemy and the occult. For our story, he is important because he was among the first to adopt a mathematical approach to science.

Working in the seventeenth century and building on the insights of Galileo, Newton uncovered his three laws of motion, of which "for every action, there is an equal and opposite reaction" is possibly the most well known. While a novice student of physics will learn the wordy description of physical laws laid down by Newton, they know that the true power is in their mathematical form. In words, Newton's second law of motion can be stated as "the rate of change of momentum of an object is proportional to the applied force and takes place in the direction of that force." In mathematics, this is reduced to the much more compact and powerful equation, $F = ma$.

Through this math, you can make predictions about the physical universe. For example, if you want to send a space probe through the solar system to explore a distant comet, you will use Newton's laws of motion and gravity to ensure that the space probe and the comet end up at the same place at the same time. But the mathematical laws are only part of the story, and it is essential to know your starting point, or in the parlance of mathematics, your "initial conditions," to make your predictions.

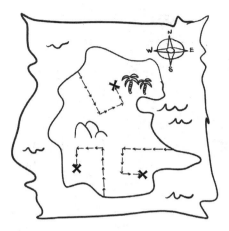

Imagine you find a treasure map that says "Walk forward five paces and turn left. Take three more steps and turn left again, then two more steps and dig." These instructions are completely useless if you don't know your initial conditions—where you are supposed to start and which direction you should be facing.

Different aspects of physics require knowledge of different initial conditions. If you want to study the motions of planets and comets around the Sun, you need to know each of their precise positions and velocities and feed that information into the mathematics. Then you can predict where the planets will be tomorrow and in the future, allowing you to make a fortune from accurate astrological predictions! You might chortle at this, but many of the motivations for accurately tracking planets across the sky over history were to enable astrology.

This "practical" use of physical laws was not only applicable to planetary motion and astrology. The field of thermodynamics grew from the need in the Industrial Revolution to understand just how much work you could get out of a machine being powered by heat. In this case, you want to know quantities such as temperature and pressure and flows of energy from one place to another. Using the mathematical laws of thermodynamics, you can calculate the efficiency of a steam engine or how long it will take an ice cube to melt in your gin and tonic.

At the end of the nineteenth century, science was coming to the realization that everything is made of atoms, and the gases that are the focus of thermodynamics are composed of an almost uncountable number of individual atoms colliding with one another and rattling around. Things like temperature and pressure are the manifestation of all this atomic jiggling. But was the devil in these details?

Here Be Demons

In theory, if we knew the precise locations of all the atoms in a particular gas as well as each of their speeds and directions, we could calculate their future paths and collisions. In that case, there would be no need for thermodynamics. But in practice, there are simply too many atoms doing their own things for us to conceivably calculate them all.

James Clerk Maxwell, the originator of the equations of electromagnetism, also pondered this question. Thinking about the motion of atoms in gases, he wondered about the action of an imaginary *demon*, a tiny creature that can see every individual atom and know their properties precisely.[1] The demon would also know the precise positions and velocities of all atoms and photons in the universe. It could then calculate the subsequent evolution of each of them.

In the nice simple universe of Newton and Einstein, the laws of physics are completely deterministic. All the demon would need to do is use all the current positions and velocities as initial conditions and then use the equations of Newton and Einstein to tell us where all the atoms and photons will be in the future.

Of course, there is no demon. And in practice, this feat would be impossible. But in theory, there is nothing in the laws of physics that forbids something that functions just as the demon would. Maxwell's demon, as a concept, has been argued about for more than a hundred and fifty years, and debates still rage on. Its implications, that thermodynamics is tied up with the concept of *information*, has proven controversial.[2] We all have an idea of what information is as a description of a thing or circumstance. Thermodynamics, on the other hand, is all about heat and energy flows. These two concepts sound so different, so distinct, that the fact that they seem to be intertwined seems, well, strange.

For some, Maxwell's demon represents a step too far, and solutions are sought to blur the link between thermodynamics and information. Many proposals to expel the demon use our now trusted tool: quantum physics.

Predictions appear much different when we consider the rules of the quantum. As Heisenberg taught us, a particle does not have a well-defined position and velocity. So this idea is already dead in the water, right? We know that the physical laws of the very small are governed by quantum mechanics, so we would need to account for this if we were going to calculate the evolution of the universe as a whole. And instead of positions and velocities, quantum mechanics encodes particle properties in the more esoteric wave function, which we'll discuss in the next section, and individual particles are not really individual but are entangled with others. So a group of individual electrons is not represented by a group of individual wave functions but a single wave function representing them all. Expanding this up to all the atoms and particles and photons in the universe, does this mean we can write down a single wave function for everything? Is the universe truly a quantum thing?

The wave function is such a tricky concept to get one's head around that physicists still argue about it today. Entrenched camps each have their preferred interpretation. They even give themselves names, like Bohmians, Everettians, QBists, and Copenhagenists.[3] But what is an interpretation of the wave function, and why does it need interpreting at all?[4] For that, we will again revisit the early twentieth century.

A Universe-Sized Wave

As Heisenberg and others were developing matrix mechanics, which led to the understanding of the uncertainty principle, Erwin Schrödinger and his colleagues were working on what seemed like a completely different calculus for quantum physics. At the time, the physics of waves was well understood and wildly popular due to the wide applications of Maxwell's equations for electromagnetic waves.

What is now known as the *Schrödinger equation* was an equation of motion for a phenomenon Schrödinger dubbed the "wave function." An equation of motion much like those of Newton and Maxwell, it followed the familiar paradigm of theoretical physics. Once the initial conditions were known, the equation did its work and predicted what this wave function would be for all future times.

The story didn't end there though. This wave wasn't like waves we are used to, carrying energy from one place to another. Nor did it somehow correspond to some physical property of the thing being studied—the location of an electron, for example. It was Max Born who demonstrated that the wave function could be used to calculate probabilities for the outcomes of measurements. The introduction of chance into the mathematics was unappealing to many, given the raging success of all the deterministic laws of physics that preceded quantum theory. You may be familiar with Einstein's lamentation, "God does not play dice!" However, it did encapsulate the same uncertainty Heisenberg found, so at least there was some consistency. In the end, Born's statistical interpretation of Schrödinger's equation was irrefutable and solidified quantum physics as a probabilistic theory.

The confusion surrounding the development of quantum theory is difficult to appreciate from today's perspective. In physics classrooms around the world, students are given the Schrödinger equation and told that it will predict the outcomes of laboratory experiments. The wave function and its equation provide a recipe to predict, control, and eventually engineer materials. The homework assignments of physics students are filled with solving this or that manifestation of Schrödinger's equation. A common example is solving the Schrödinger equation for the hydrogen atom. The solution, which accurately explains the energy levels inside the hydrogen atom, is built of complex functions called spherical harmonics that produce the beautiful *orbital* shapes seen in physics and chemistry textbooks. Students are told that these are some sort of fuzzy representation of where the electron *is*, and it is left at that.

For many decades, this metaphysical question of what exactly this wave function is has been responded to with the now infamous answer, *shut up and calculate*. Thus, by now, the prevailing attitude among a vast majority of physicists is that the wave function is a calculational tool only. But the curious mind is not satisfied so easily. To many practicing quantum physicists, there are two modes of operation. Given a well-defined problem, the quantum physicist will indeed shut up and calculate. However, when the calculations are done and the thinking begins, the quantum physicist is never entirely satisfied with even their own understanding of the wave function. Painted in broad brushstrokes, the question can be raised in several ways. *What does quantum physics tell us about reality? What part of reality does the wave function correspond to? Is the probability in quantum theory part of reality or our knowledge of it?*

Interpretations of the wave function are closely related to interpretations of probability, which can be neatly divided into two camps of thinkers. The first group considers probabilities to be *objective*—they are real. For example, when we say there is a 50:50 chance a coin will come up heads, that chance is a real property of the coin, often called its bias. This is the intuitively obvious way to think about chance for a pit boss at the casino, who is trying desperately to identify those coins or dice that have been tampered with to weight the odds in the gambler's favor. For a large part of the twentieth century, mathematicians and statisticians held this view as well. This in turn had great influence on the physicists and philosophers of the time.

The second camp considers probabilities to be *subjective*—they exist only in the mind of the observer. In the case of the coin, it is I who assigns 50:50 probability to heads, not the coin calling out its own unbiasedness. I don't know whether the coin is fair or not, so what choice do I have but to assign 50:50 to the possible outcomes of a toss? To those holding the subjective view, probabilities are just numbers representing the private expectations of people. Though this interpretation has steadily been gaining popularity among both statisticians and physicists in recent decades, there is still no consensus on it.

Quantum Interpreters

These interpretations of probability are echoed in quantum physics. In the context of the wave function, one camp adheres to the view that it corresponds directly to reality. The wave function is considered a real part of the world for them. The other camp views the wave function as subjective. A wave function is something personal to a scientist, who uses it for their calculations and nothing more. There is no right answer here. However, if your inclination is toward an objective wave function, then you might also be seduced by the idea of a *universal wave function*. For if the wave function corresponds to reality, then the reverse should also be true. That is, all of reality—the entire universe—should possess a wave function.

The idea of a universal wave function is not new, first appearing in the PhD thesis of Hugh Everett III in 1956. Everett developed the strange consequences of this idea. In particular, it led him to the infamous *many-worlds interpretation*, which we will get to shortly.[5] Others, including physicists like Stephen Hawking, took the idea of the many worlds seriously.

This wave function of the universe obeys the Schrödinger equation, as all wave functions must for quantum physics to be valid. At each point in time, the equation tells us the wave function of the entire universe. Running the equation backward in time, we end up with the wave function at time zero. This must be the initial state of the universe. Wave functions can tell us all sorts of useful properties of things. We have already discussed vacuum fluctuations and exotic phase transitions, such as the production of inflatons driving inflation, as properties of this initial quantum wave function of the universe.

The issue with any interpretation of wave functions is the role of the scientist, the so-called *observer*. The rules of quantum physics—honed to be the most precise scientific theory ever devised—demand that the Schrödinger equation stop when the observer acts. When the observer acts, it's as if time is reset. The wave function changes violently and instantly—a process called collapse. It is often said that the wave function encodes the idea that everything that *can* happen *does* happen. Yet we, the observers, only see one possibility. (The coin comes up heads *or* tails, not both.) We collapse the wave function. How can it be, then, that the entire universe is described by a wave function if the action of a single observer can change it? For that matter, who—or what—is allowed to be an observer? A scientist? A rat? A politician?

Ignoring the problem of the mind, or consciousness, everyone agrees that humans are made of physical stuff. Thus, we, too, should be describable by quantum physics. Indeed, we ought to be part of the variables going into the universal wave function. But it doesn't *appear* that way to us. Enter the most controversial idea within the scientific field and the one most beloved outside it. The many-worlds interpretation, initially proposed by Hugh Everett III, is the one idea from quantum physics that storytellers and filmmakers have wholeheartedly adopted. Who does not love a story where the protagonist ends up in a parallel universe where the Allies lost World War II or the British won the American War of Independence? Apparently, historians dislike counterfactual history, but science fiction fans love it!

Within the physics community, however, the many-worlds interpretation is the cause of debates about as heated as academia can get. The many-worlds theory claims that there is only one wave function, the universal one, which is always evolving according to the Schrödinger equation. Everything that can happen does happen. Since the wave function corresponds to reality, and it seems to encode multiple distinct possible realities, those realities must all exist, so the theory goes. Many realities, many worlds.

In the many realities of the many-worlds interpretation, there are observers with quite distinct perceptions. You could see the coin land heads, or you could see it land tails. According to the many-worlds theory, both are equally *real*. From your perspective—say you are the observer who saw heads—the coin landing heads is the only reality. But the many-worlds interpretation suggests that another observer exists, identical to you in all respects except for the fact that he or she saw the coin land tails. Both are realities that play out in parallel, part of the large, evolving universal wave function.

Before we close this chapter, we're sure that the reader has raised their eyebrows a little with this concept of a single wave function for the entire universe. We have definitely strayed from what some would see as robust science into the realm of scientific speculation. Some would even suggest we've forayed into scientific daydreaming. But in reality, we are hitting the murky interface of the language of quantum mechanics and that of general relativity. We don't know if we can truly describe the universe in terms of a wave function, but it is a speculative idea.

At this point, it's time to leave the idea of a universal wave function behind and step into the apparently endless future that awaits the universe. This universe of tomorrow will be very different from the universe of today, and we'll see that we have to rely on more speculative unions between quantum mechanics and general relativity to try to imagine what it might be like. In the future, gravity and the other forces will still be vying for dominance in shaping the universe. It's time to see what an interesting and quirky universe it could eventually be!

PART 3

The
QUANTUM
of COSMOS
FUTURE

Why don't all dead stars become black holes?

Through the preceding chapters, we've come to realize that stars have lives; they are born, they live, and they die. How a star dies depends upon its mass, as this dictates the squeezing due to gravity and consequently the rate of nuclear reactions at its core. This means that some stars *can* end their life in a bang, but for many, it is more of a whimper.[1]

Let's look again at the most massive stars. As we have seen, these stars can end their lives spectacularly in immense supernova explosions that can be seen across the universe. Here, an entire star can be ripped apart by the push of the uncountable ghostly neutrinos. Let's review what actually happens inside the star.

As a massive star ages, nuclear burning in its core continues until iron is produced. But iron is different from all the other elements that came before it in the star's life cycle, and fusing iron into heavier elements actually sucks in energy rather than emitting it. Suddenly, the nuclear fires in the star's core are switched off, and the outward push from radiation vanishes. There is nothing to prevent gravity's inexorable squeeze, and the star collapses down on itself. The density soars, the temperature soars, and as iron gets forced into heavier elements, the end burst of neutrinos is released, and the star explodes outward.

Well, not all of the star. As the density increases in the very core, so does the pull of gravity, accelerating further collapse. At some moment, a critical point is passed and nothing can stop the inward pull of gravity, and a black hole, usually several times more massive than our own Sun, is formed. This black hole is the remnant of the massive star, surrounded by an expanding and fading shell of debris from the explosion.

For slightly less massive stars, the process is very similar, but while the density and gravity can soar in the collapsing core, they never reach the critical point to form a black hole. The collapse can be halted! But this only happens after electrons are squeezed into atomic nuclei and into protons, creating neutrons. The resulting star consists entirely of neutrons. Such a neutron star is an extremely weird place, like nothing on Earth.

The death of a less massive star, like our Sun, is less dramatic. The Sun is roughly halfway through its expected eleven-billion-year lifetime, and as it approaches its end of days, it steadily changes its internal consistency as heavier and heavier elements are forged.

The Sun doesn't have the gravitational squeeze to create elements like iron, but its changing internal structure will cause it to swell to an immense size, becoming a red giant star, ultimately consuming Mars. Its unstable nuclear reactions will cause it to pulse more and more violently, blowing off its outer layers and leaving little but the stellar core.

This dead core of the star, known as a white dwarf, will be the extremely hot and dense remnant of the stellar heart. It will be about the size of the Earth, with the mass of about the Sun, but it will no longer be able to sustain any nuclear reactions. Its heat will provide an outward pressure that can hold gravity at bay, at least for a while. While born hot, this white dwarf star will eventually begin to cool down and, over many billions of years, will fade to be as cool as the background universe, a truly dead star known as a black dwarf. The time needed for a white dwarf to cool down to the background temperature of the universe is immense, many times longer than the current age of the universe, so no true black dwarfs may yet exist, but they will be there in abundance in the future universe.

We should complete this story and consider the lowest mass stars. These, the red dwarfs, are the most numerous stars in the universe today, and their ends will be completely undramatic. Due to their low masses, their nuclear reactions are sedate, burning slowly and steadily for more than a hundred trillion years. But once the nuclear fuel, the hydrogen in their cores, is exhausted, there is little more these tiny stars can do. They simply go out and fade into the darkness. Once dark, dead red dwarfs will still retain a bit of heat, providing a little pressure to prevent collapse. But they, too, will eventually cool, losing all energy into the darkness of the universe.

But there is something puzzling about these dead stellar remnants. After the outward push from nuclear reactions or the pressure of heat is gone, why don't they all succumb to gravity's immense squeezing and collapse into black holes?

You might wonder if they are like the Earth, which is not collapsing even though it is not supported by nuclear reactions at its core. In the Earth, it is the electromagnetic attraction and repulsion of atoms that can overcome gravity's squeezing, providing enough push to prevent collapse. But dead stars are much more massive than the Earth, with much stronger gravity overcoming the pressure provided by electromagnetism. So where does the force that prevents their gravitational demise come from?

A Quantum Lifeline

Think back to when we were pondering the early universe and discussing the deuterium bottleneck, the roadblock holding up the creation of elements as the universe cooled. Remember that deuterium is the basic two-nucleon building block of matter, as the other two-nucleon possibilities—the diproton and dineutron—are unstable and immediately fall apart. The reason has to do with their spins and the nuclear force. The final ingredient is the Pauli exclusion principle, which demands that no two fermions can share the same state. In the parlance of the previous section, fermions cannot have identical wave functions.

The counterpart to a fermion is a boson. Bosons include the force-carrying particles, such as photons, but also higher-mass composite particles, including our friends deuterium and helium-4. Bosons are not constrained by the Pauli exclusion principle and can thus occupy the same quantum state. Photons, which are a member of the boson family, can bunch together in laser pulses—for example, think laser eye surgery, not *Star Wars*. The most powerful—highest energy—laser pulses last ten nanoseconds and contain as many photons as there are atoms in your body. When bosons get together, you cannot think of them as individual entities anymore. There is only *one* wave function that describes *all* of them.

As exotic as it might sound, condensed, massive bosons are now routinely created in physics laboratories by cooling down gasses of these particles to near absolute zero. From there, all sorts of wild things can happen, such as superconductivity (electric current without resistance) and superfluidity (motion without viscosity). But we don't want to talk about bosons, since most matter is made of fermions. Fermions do not condense to occupy the same wave function, a consequence of the exclusion principle. Pauli introduced his idea as an accounting mechanism to explain why electrons seems to arrange their energies in the odd way they do, as is easiest to see in the patterns in the periodic table. The idea was quickly elevated to a *principle* from which one could derive the arrangement of electrons in atoms once spin was introduced.

At the time, it was understood that higher-energy electrons spent their time farther from the nuclei of atoms. An atom with either high energy or a lot of electrons thus occupies a lot of volume, relatively speaking. The fact that higher-energy electrons occupy more volume had already been measured in experiments. Shortly after Pauli introduced the idea, Paul Ehrenfest pointed out that going in the opposite direction has interesting consequences. If we try to condense atoms, the electrons get pushed closer to the nuclei to occupy less volume. But the Pauli exclusion principle forbids this, as these electrons cannot share wave functions. This application of Pauli's idea demonstrates why mass has volume at all.

We rudely reminded you of your high school chemistry home-work without warning in a previous chapter. Perhaps, though, you preferred filling the orbitals of atoms with electrons to dissecting frogs. The exercise of course turned into one of rote memoriza-tion, though a few mnemonics exist. The key task was to remember when to pair electrons with spin down with electrons with spin up, a task disguising the Pauli exclusion principle. In the lowest energy level, 1s, only two electrons were allowed—one would have spin up and the other spin down. Yet spin, you'll recall, is an internal degree of freedom not (yet) contributing to the overall energy of the electrons in an atom. Two electrons can have the same lowest possible energy so long as they have opposite spin. In other words, fermions can have the same energy without sharing wave functions. The terminology for this in physics is *degeneracy*.

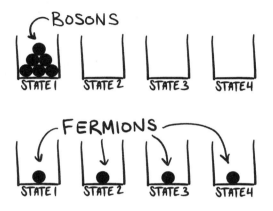

The Pauli exclusion principle requires that no two fermions share the same set of values for all observables. However, they can share all but one value. The most obvious values are imported from our classical intuition—things like position, speed, energy, etc. Fermions are free to share all these so long as their internal quantum degrees of freedom are different. This is why two or more electrons in an atom can share the same energy level. The lower the energy, the less degenerate it is. Since the lowest energy level in an atom, 1s, can support only two electrons (one spin up, one spin down), the remainder of the electrons must possess higher energy. Thus, even cooling atoms to absolute zero results in electrons with high energy. A cloud of electrons—called a Fermi gas on account of them being fermions—resists compression precisely because of the Pauli exclusion principle. A resistance to compression is otherwise known as *pressure*. To distinguish this from the pressure caused by heat in a normal gas, we call it *degeneracy pressure*.

Two masses—be they atoms or planets—are attracted to each other through the force of gravity. When one of the masses is very big compared to the other, we often think of that mass as being fixed in place while the other either falls into it, escapes off to infinity, or orbits around it. For example, the Sun accounts for 99.9 percent of the mass of the entire solar system. This means the center of gravity of the solar system can be taken to be the center of the Sun, for all practical purposes. The regular motion of the planets around the Sun is a result of the enormous size of the Sun compared to that of the planets. When the masses are of comparable size, the motion becomes more complicated. Pluto's moon Charon, for example, does not orbit Pluto as our Moon orbits Earth—Charon and Pluto orbit the center of gravity of both their masses, which does not lie within Pluto. They are in a perpetual cosmic dance around each other.

More objects have very complicated—technically *chaotic*—behavior. However, even if their trajectories are not simple circles or ellipses, they all still orbit a center of gravity, and as they lose energy, they fall toward that center of gravity. This is how things get compacted—how stars and planets form. Gravity is always pulling things together. But if gravity were the only force, everything would end up as a single gargantuan mass in an infinitely small point. For a star like the Sun, something must counterbalance the force of gravity to give it shape. As the Sun burns hydrogen into helium, it is the outward pressure from the radiation flowing from its core, but once this is gone, gravity will compress the core until degeneracy pressure stops it. This is the pressure that stops dead stars from collapsing into black holes!

Cosmic Weight Watchers

Once our Sun becomes a white dwarf, it will be a Fermi gas of electrons with helium and carbon nuclei swimming about. This will support this remnant of the Sun as it loses its heat and cools into a black dwarf and beyond.

For stars larger than the Sun, the extra mass causes more force from gravity, compressing the white dwarf to an even smaller size and forcing the electrons to have even more energy. But there is a limit. Electrons cannot have so much kinetic energy that their speed would be faster than light—the speed limit set by special relativity.

Physicist Subrahmanyan Chandrasekhar calculated what this limit would be in terms of mass.[2] The Chandrasekhar limit, as it is now known, is about 1.4 times the mass of our own Sun. For a star to have left a core of this size, it would have been about eight solar masses during its lifetime as a hydrogen-burning star. A star larger than that suffers an even stranger fate.

With sufficient force from gravity, the electrons in the core of a dying star are forced into nearby nuclei where they react with protons. This process, called *electron capture*, creates neutrons and neutrinos. The neutrinos radiate away, leaving a stellar object consisting only of neutrons—a neutron star. But without electron degeneracy pressure, how does such an object remain stable?

A partial answer is exactly the same one as we had for electrons, since neutrons are also fermions with spin ½. However, neutrons are not fundamental and can be chopped into smaller pieces, with every neutron composed of three quarks. So the answer is not simply neutron degeneracy pressure. Additional nuclear forces are at play but not yet fully understood. As with all good science, there are mysteries yet to be solved.

Recent measurements of a neutron star using gravitational wave astronomy weighed it at about two solar masses. Current theories suggest that the most massive neutron stars can be three solar masses, as beyond this, not even degeneracy pressure can save the star from the force of gravity. What happens when the gravitational force is so strong as to force the neutrons past known relativistic limits? That results in what are the most enigmatic objects in the cosmos: black holes.

Before we explore the influence of quantum mechanics on black holes, we have to come clean on a few things. We started this section discussing the stability of black dwarfs, the cold remnants of dead stars that are expected to fill the future universe. These don't collapse due to the degeneracy pressure explained by quantum physics and the Pauli exclusion principle. Electrons, like all fermions, just can't bunch up in the same place—you can squeeze and squeeze, but degeneracy pressure will resist that squeezing. It seems our future universe, filled with dead star hearts of degenerate matter, will be a very strange place indeed.

But this is not the full story. While degeneracy pressure will become extremely important in the future universe, its influences are already felt all over cosmic history. There are already neutron stars in the universe, left over from earlier generations of stars that lived and died, many before the Sun was even born.

There is a one last twist to our story. We've mentioned that red dwarfs are the smallest stars, and they have a mass about a tenth the mass of the Sun. But why are there no smaller stars than them? There are small clumps of gas that can collapse, and gravitational squeezing heats their cores, but as the matter gets denser and denser, degeneracy pressure rapidly comes to dominate, preventing further collapse. The cores of these stillborn stars are never hot and dense enough for nuclear fusion to ignite, and they, the brown dwarfs, are destined to roam the cosmos in the shadows.

In fact, there is such a failed star in our very own solar system—the planet Jupiter. It had a different formation than a brown dwarf, but the physics is the same. The heart of Jupiter is about half the density of the center of the Sun but more than six hundred times cooler. The conditions are not extreme enough for nuclear fires, but its core cannot collapse any further due to the effects of quantum mechanics.

Stop and think about that when you spot this majestic planet on a cool, clear night.

Will matter last forever?

In a few hundred trillion years, the last stars will have burned out. The universe will be plunged back into darkness. It will be filled with dead stars, radiating their remaining heat into the void as they cool toward absolute zero. Perhaps this is it for the universe—its end state in which it will exist for the rest of eternity. But as we will see, the action of quantum mechanics means that matter itself might eventually melt away into the darkness.

Life is a permanent battle against decay, and without constant repair and upkeep, everything breaks down, be it your car, your house, or even your body. Decay is inevitable. But to the universe, at its most basic level, decay is an illusion.

When food rots or iron rusts, chemical bonds bind and break, but the atoms that form the molecules and crystals that underlie matter remain unchanged. If we continue to break things down, to pull things apart, eventually all we will have are the individual atoms that make up all the matter in the universe.

While atoms appear to be permanent, we know this is not truly the case. The elements were built up in the early universe and in the hearts of stars, and some of them can break down through the action of radioactivity. Some atoms indeed appear to be stable, resistant to the actions of radioactivity, and will last into the long, dark future of the universe ahead.

But what of the protons and neutrons that make up the nuclei of atoms? How stable are these? It would seem that these too must be stable given that some atoms, containing protons and neutrons in their nuclei, appear to be completely stable. But if we take a single neutron, leaving it on its own, in about fifteen minutes, it will be gone.

A neutron can decay because it is has slightly more mass than a proton. The mass difference is not much, only 0.1 percent, but it means there is enough energy in the neutron to be converted into a proton, an electron, and an almost massless neutrino with a little bit left over (most of which goes into the motions of the electron and neutrino). This, of course, is the famous law of the conservation of energy: energy cannot be created or destroyed, just transferred from one kind to another. Given this, there must be the same amount of energy before and after the neutron decay, no more, no less.

What about a single proton? Does it decay like the neutron?

Given that it is lighter, the proton cannot decay into a neutron, as this would violate the conservation of energy; there is simply not enough energy in a proton to be converted into a neutron.

That is fine, you might say. Why doesn't the proton just decay into something else, something with less mass so the decay will not violate the conservation of energy? The proton could decay into a thousand electrons, and there would still be plenty of energy left over to make even more electrons. But when scientists watch and wait, observing individual protons, they do not decay into a thousand electrons or a hundred or even ten, all possibilities allowed by the law of conservation of energy.

Quantum Bookkeeping

Something else is going on here, preventing protons from decaying. In fact, other conservation laws are in play, conservation laws written with the laws of quantum mechanics. In particular, there is a bit of quantum bookkeeping going on, where we have to keep track of a property known as *baryon number*. As this is a conserved quantity, we must observe the same baryon number before and after a reaction.

While it sounds like strange physicists' jargon, a baryon number is quite simple—the baryon number of a system of particles is the number of quarks minus the number of antiquarks, then divided by three. Why three? Mostly convenience. The vast majority of matter in the universe is made of protons and neutrons. Each proton and neutron is made of three quarks, so each has baryon number 1 (3 divided by 3). Quarks, therefore, have baryon number ⅓. All other fundamental particles are given a baryon number of 0. The antiparticles are given the negative baryon number of the particle cousins. For example, an antiquark has baryon number −⅓, and an antiproton has baryon number −1.

The entire universe has a baryon number as well. It's big. It has also remained unchanged since the early universe. We've already traveled back to this time when we asked *where did all this matter—*err...baryons—*come from?* We didn't give it a name at the time, but the moment when matter began to dominate antimatter—leaving more quarks than antiquarks—is called *baryogenesis.* Now that sounds epic! Of course, we don't know exactly what happened further back than we can see—we would love a way to directly test theories about the early universe! But we suspect that the symmetry between matter and antimatter was broken, which leaves us with more quarks than antiquarks and a large net baryon number. And as if that spontaneous breaking of symmetry wasn't strange enough, today we seem to have recovered that symmetry, as no experiment has ever uncovered evidence for the nonconservation of baryon number. The question is, how long will this symmetry reign?

Just as our familiar notions of energy and charge are conserved, we might also accept that mass is conserved. In school textbooks, we would simply be told that *matter cannot be created nor destroyed.*

Where does this theory come from? Why, quantum physics of course! Within the four fundamental forces and the standard model that describes them, there is no way for particles to interact to change the baryon number of the system. In other words, as Emmy Noether taught us earlier, there is a symmetry within the mathematical laws that prevents changes of baryon number—baryon number is *conserved*. This reemphasizes the conundrum of baryogenesis—it's not in our current physical laws, which are built on the conservation of baryon number.

Conservation of baryon number is immensely useful. It is used for hunting for new particles in billion-dollar physics experiments, allowing us to uncover the vast array of particles spat out in collisions, as well as for checking your quantum physics homework assignments. Let's see if *you* can ace this advanced quantum physics quiz. Ready? Can the interaction proton + neutron $\rightarrow$ proton + proton + antiproton occur? Hmm. We start with a positive charge and end with a positive charge, so that seems okay. But we start with 2 baryons and end with 2 baryons and −1 baryons, totaling 1 baryon. So baryon number is not conserved. The answer then is no, this interaction cannot occur.

Now on to the second question. Can the interaction proton + proton $\rightarrow$ proton + proton + proton + antiproton occur? Charge is conserved—check. Also, the baryon number now remains at the value of 2. So it seems that yes, this interaction can occur. Indeed, this *proton-antiproton pair production* can be observed when the two incoming protons are smashed together with enough energy. But in this interaction, the number of protons has *increased*. If we are looking for the decay of protons, we seek an interaction where the number of protons *decreases*. But the proton is the least energetic of baryons. That means that decay to something more energetically favorable would have to change the baryon number—forbidden!

The proton is protected by symmetry. Or to put it in a way that is fit for a discussion about the long, lonely death of the universe, the proton is doomed to an eternal life. Maybe.

The Inevitable: Death, Taxes, and Decay

Before we get to how a proton might decay, perhaps it is worth stepping back and asking what decay even is or, more to the point, why it happens. Though our current understanding uses quantum physics in its full glory, it wasn't always so. Decay means to deteriorate, decompose, or rot. In the early 1900s, the systematic study of what they called "radioactivity" led Marie Skłodowska Curie and her husband, Pierre, to find that the elements themselves are subject to decay.[1] Some elements emitted radiation but less and less over time—whatever the source was, it *decayed*. It was discovered that such decay resulted in one element changing into another.

In quantum physics, decay now means any process in which a high-energy state of a system moves to a low-energy state. Phrases you may have heard include "the atom has decayed," which means it now occupies a lower energy state. But we know that energy is always conserved, so that lost energy must go somewhere. This is why all decay events result in some emission of energy. If a lower energy state is available, the system is *unstable*. A subtle distinction is made when the system in question is an unstable nucleus, such as that of the infamous element uranium. We call the process in this case *radioactive decay*, because the emitted radiation is high energy and potentially dangerous. It is often called *ionizing radiation* because it contains enough energy to strip electrons from the material it passes through, which happens to have a negative effect on living cells in high doses.

How much radiation a sample emits obviously will depend on how much of the stuff is sitting there. More of the sample means more decay, and less of the sample means less decay. Of course, we can't use this crude rule of thumb to make predictions. Luckily, the precise mathematical rule is only slightly more complicated: how much of a sample will decay in some instant of time is proportional to how much matter you have. This observation is actually quite remarkable. It means that over a fixed amount of time, the sample will decay by the same fixed proportion. For example, if it takes sixteen hundred years for a sample of radioactive radium to be reduced to half, it will take sixteen hundred more years for what's remaining to again be reduced to half, and so on. In this case, sixteen hundred years is called the *half-life* of radium.

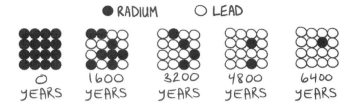

The half-life is a fundamental property of an element and is different for each type of element. Some elements, like hydrogen-7 (hydrogen with six neutrons), have a half-life of *yocto*seconds (a trillionth of a trillionth of a second), and others, like lead-204, have a half-life of *yotta*seconds (trillions upon trillions of seconds). This defines the stability of the element—a half-life of a yoctosecond means a very unstable element, whereas a half-life of a yottasecond (being millions of times the age of the universe, by the way) means the element is stable for all practical purposes.

Now, you should be wondering how we would even know that the lifetime of an atom is millions of times the age of the universe. Did someone wait around that long to measure it? Of course not, but here is where quantum uncertainty comes to the rescue yet again. Let's go back to radium and its sixteen-hundred-year half-life. If we have 1 gram of pure radium (radium-226 specifically), then after sixteen hundred years, we will have ½ gram of radium due to decay. But it's not the case that once the clock strikes sixteen hundred years, half of the radium decays. The individual atoms each have a small chance of decaying at any given second of time. After a second has passed, some have decayed, while the rest retain that same small chance in the next second. After fifty billion seconds (about sixteen hundred years), roughly half the atoms have decayed. Even though this sounds like a long time, we must remember that there are over a sextillion individual radium atoms in each gram. With a half-life of sixteen hundred years, this translates to thirty-seven billion decay events every second! To put that in perspective, your body, which contains radioactive potassium, produces about eight thousand decay events per second. (Yep, you are radioactive, as is your computer, your dog, and any piece of matter you can think of.)

The half-life is a number that tells us something about probability, not something real or physical. A given atom has a probability, no matter how rare, of decaying at any instant. If there are enough chances for a rare event, it will eventually happen. That's why spent uranium fuel rods from nuclear power stations are dangerous now and will remain dangerous long after humans have disappeared. So the question of whether protons last forever is really the question of whether there is a lower energy state they can occupy and what the half-life of decaying there is.

Protons, being baryons, must decay into other baryons—that is if baryon number is to be conserved. But like the other anomalies we have encountered, breaking this symmetry is a fun exercise for physicists. Sometimes, physicists break the mathematics of the current standard model to allow for proton decay, and sometimes completely different models meant to overthrow the standard model allow for it. There is an endless stream of proposals, but there is a problem. With every particle physics experiment, the standard model is vindicated, explaining the electrons, protons, and assortment of other particles that are spat out. With no cracks in its mathematics, many of the alternative proposals go into the garbage, while others sit and wait for the standard model to fail.

Scientists have continued to search hard for any signs of protons decaying. Of course, given the immense potential lifetime of a proton, there is no point isolating a single proton in a laboratory and simply watching it. To increase the odds of finding one in a state of decay, scientists watch a lot of protons at the same time.

Putting a Best Before Date on Eternity

Remember that the lifetime of a quantum thing, such as a particle, is a statistical question. If we say that the particle lifetime is one year, this means there is a 50 percent chance it will decay in one year. If it hasn't decayed in year one, it has 50 percent chance of decaying in year two, and so on.

So while the lifetime of a proton may be staggeringly long, there is a chance—a small chance—that an individual proton might exist for only five minutes before decaying. If you watch a huge number of protons, say a swimming pool–sized vat of atoms in molecules, and know what the signature of a decaying proton looks like, namely the mysterious appearance of a fast-moving positron and other stuff, then you have a chance of witnessing one in decay. This is precisely what physicists have done, but so far, they have not seen the "smoking gun" of proton decay.

Scientists are undaunted by the current lack of experimental evidence for proton decay and think it's just a matter of time (pun intended) before it will be observed. But what does evidence of a decaying proton mean for the future universe?

After a few hundred trillion years, all the stars will have exhausted their nuclear fuel. Without starlight, the universe will descend into darkness, into a night that will last forever. The dead stars will exist in the darkness, cooling and fading into the background of the universe. But as we push out into timescales of 10^{40} years, a mighty long time after the last star has been extinguished, the decay of protons will start to be felt.

As protons decay, matter will start to melt away. In the darkness, the black dwarfs will evaporate into simple particles and light. After a few proton lifetimes, they will have dissolved completely into the nothingness. After protons have decayed away, the epoch of matter will be truly over, and there will be nothing but elementary particles and black holes inhabiting the universe.

Of course, proton decay is not certain. Physicists think there are good reasons for protons not to be stable for eternity, related to the holes in our theories of quantum physics. But even if protons do not decay, it does not mean the universe will never change. Inside the dead hearts of stars, atoms will be squeezed very tightly together. This close proximity will very, very rarely allow for *cold fusion*, where quantum tunneling will meld atoms into new elements. This process will continue, very, very slowly, until all atoms are squeezed into a strongly bound atomic nucleus: iron-56. Without proton decay, after an unimaginably long period of time, around $10^{1,500}$ years, all the matter in the universe will have fused into this eternal state of iron.

Do black holes last forever?

If protons do decay and once all the dead stars have melted away, other masses will still lurk in the darkness. Black holes, many formed from the collapse of stars in supernovae, will be impervious to the actions of proton decay, their matter locked up in their infinitely dense cores, a point mass that is known as the *singularity*, the name given to places in physical theories where infinity occurs.

Physicists don't think the cores of black holes are true singularities. In fact, physicists think that the infinities of singularities have no place in theories about the actual universe, and something prevents them from existing. For black holes, it is speculated that the actions of quantum forces will ultimately prevent the formation of a point of infinite density, but the immense gravitational pull of the black hole will remain. Stuff that falls into a black hole is destined to remain there, and black holes will remain long after the last star has dissolved into the blackness.

But will black holes last forever? Is the distant future of our universe destined to be an ever-thinning sea of elementary particles with a scattering of black holes? Within Einstein's general theory, black holes are truly eternal, able to grow by eating material. As matter falling into a black hole is on a one-way journey and can never get out again, black holes can never shrink.

At least that's what everyone thought until the 1960s. It was then that a young researcher named Stephen Hawking began to ponder the nature of black holes. His focus was not the singularity, which is a mathematically confusing place to explore, but a region around the singularity known as the *event horizon*.

The existence of the event horizon was known from the earliest mathematical explorations of black holes. Located at a particular distance from the singularity, the event horizon represents the "no way back" boundary. Things that cross the event horizon are compelled, by gravity, to fall into the singularity. Nothing can stop this from happening. No amount of struggling or rocket power can prevent this fall to the center once you are below the horizon.

Strange things can happen at the event horizon. The bending of space and time means that light struggles to escape from just outside the event horizon, but at the horizon itself, light can be held, motionless. For objects that fall into a black hole, an image of them, representing the last emission of light as it crossed into the abyss, is imprinted on the horizon.

Hawking, an expert on relativity, wondered what would happen to quantum mechanical processes occurring at the event horizon. What would be the impact of a one-way boundary? Hawking's conclusion was extremely surprising, finding that quantum mechanics means that black holes radiate, converting the mass in the singularity into a faint glow at the event horizon. And the impact of this *Hawking radiation* on the long-term stability of black holes is stark. But to understand Hawking radiation, we first need to talk about the thorniest concept to come out of quantum physics: entanglement.

Where there is smoke, there is fire. Where there is rain, there are clouds. Both of these examples are only *correlations*. It would seem even scientists constantly need to be told that correlation does not imply causation. For example, clouds do not always imply rain, and fire does not always imply smoke (it depends on what is burning). And in some correlations, neither event causes the other. A famous example is the fact that the number of crimes in a city is higher when the number of police officers is higher. Do police cause crime, or does crime cause police? In fact, it is neither. The population of the city accounts for both. A larger city has both more crime and more police. Population in this case is called a *common cause*.

In the theory of causation, there are only three possibilities when it comes to a correlation between two events. Either the first event causes the second, the second causes the first, or a third, unseen event caused both. But wait, how could the second cause the first—is the future influencing the past? No. We don't need to see the events with our own eyes in the order in which they are caused. For example, we often see smoke before we see fire, but fire clearly caused the smoke. The only other important point about causes is taken from physics: the chain of events, from cause to event, must obey the laws of relativity. In other words, causes, like all transmitted information, are limited to traveling at the speed of light or lower.

Playing Games

Let's play a hypothetical game. Suppose someone takes a pair of gloves and places each one in a separate box. The boxes are unmarked, and only the person who placed them in knows which is which. They give one to you and one to your friend and send you each off to different parts of the country. You have a box in Los Angeles, and your friend has a box in New York. Neither of you knows which glove you have. You know it is either a right-hand or left-hand glove, but it is a coin toss as to which happens to be in your box. You open the box—you have the left-hand glove! And now, all of a sudden, you know exactly what your friend will find in their box.

This little game could be repeated over and over. Each time, you find either a right-hand or left-hand glove. Either case happens about 50 percent of the time, but neither you nor your friend knows in each particular trial which case it will be. What remains is the correlation—in this case a perfect correlation—between the contents of each box.

Opening your box to reveal a right-hand glove does not *cause* your friend's box to contain a left-hand glove. Likewise, your friend's actions and the contents of their box do not cause the opposite handed glove to appear in your box. In this case, there is a common cause—the person who separated the gloves in the first place! That person knew who would find which glove the entire time, even if they were sending you off with different handed gloves each time. In the parlance of our previous discussions, the result was determined, and in all such cases, we can trace back the chain of events to a common cause.

Suppose instead of gloves, someone put electrons in each box. You have an electron in your box, and your friend has one in theirs. You take the boxes and go your separate ways. In Los Angeles, you open the box and find the spin of your electron is up. Your friend opens their box at the exact same time and finds the spin of their electron is down. If you repeat this game many times, you find— just as was the case for the gloves—sometimes your electron has spin up, and sometimes it has spin down. The sequence of ups and downs occurs more or less randomly. But you and your friend always find electrons with opposite spin. Okay, no big deal—just apply the same logic that we applied to the gloves in the boxes. That is, the person who put the electrons in the boxes arranged for the situation to unfold exactly as it did.

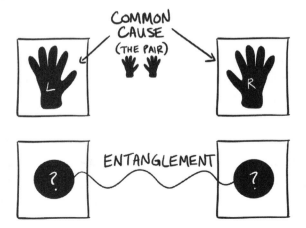

Of course, in nature and in physics experiments, we aren't playing games like this, but it is still a good analogy. The role of the person putting the electrons in boxes could be played by atoms or even distant stars. The essential point is that there is a common cause for the correlations observed even at the seemingly random quantum level. This, Einstein argued, must be true to avoid what he called "spooky action at a distance," meaning that one electron seems to influence the other from Los Angeles to New York. Although Einstein thought there must be a common cause, it wasn't at all clear what the common cause might be—it did not appear in the lab or in the mathematics. The common causes of quantum correlations thus came to be called *hidden variables*.

But here is where things get really spooky. Hidden variables do not exist. There is no common cause for quantum correlations. In our game, the person who put the electrons in the boxes did not—in fact, could not—know what you and your friend would find upon opening. In other words, it is possible to create correlated events that could not have been predetermined. We still know the electrons are correlated before their spins are measured, and this correlation is called *entanglement*. As far as the deep concepts in quantum physics are concerned, quantum entanglement is probably the most recent to be fully appreciated.

It wasn't until the late 1960s that John Bell proposed an

experimental test to prove that nature indeed behaves in this way. And it wasn't until the most recent decade that conclusive experimental evidence was provided, although it was done using photons, not electrons. While our cute little game suggests otherwise, manipulating and maintaining the states of entangled photons across large distances in the real world is an incredible challenge. But it is done, and we now routinely manufacture correlated events in the world that have no cause at all.

Perhaps this is actually not surprising if you take the uncertainty principle to heart. The uncertainty principle taught us that some properties of quantum particles cannot be defined, that they do not exist before being measured. This certainly sounds like a handicap both scientifically and technologically speaking, but it is not so! Entanglement is the basis for provably secure cryptography, a secret code that cannot be broken, and it can even teach us a thing or two about black holes! Before we get to that, though, we want you to do something... Toss an encyclopedia into a fire.

Black Hole Firewalls

No, don't really burn your encyclopedia. Scientists would never condone burning books, but this is the best analogy we have. The encyclopedia, full of all sorts of good information, will of course burn. And after the pages burn up, all the information has been lost—or has it? In principle, we could collect all the smoke and ashes and meticulously piece them back together to recover the encyclopedia. This is yet another conservation law: the *conservation of information*. This law stems from the basic time reversal symmetry of the equations of quantum physics. Even if you tossed the encyclopedia into a black hole, in principle, the information is still there, somewhere. In any case, there are many copies of the encyclopedia, so no real information has been lost.

Consider the tragic scenario of your friend falling into a black hole with their unopened box containing one glove. We did not get to see what was inside, so that information is lost, right? No. The person who put the gloves in the boxes still possesses it. However, if it were the box with the electron in it, we have a problem. Since it is impossible for anyone to know what the state of the electron that fell into the black hole was, it seems the information is lost to the universe! This is the infamous *black hole information loss paradox*, where a black hole has no detailed memory of what has fallen in.[1] And it gets worse.

Suppose the entangled pair of particles in question is one of the particle-antiparticle vacuum fluctuations we met way back at the beginning of the universe. Further suppose the antiparticle falls into the black hole and the particle does not. The antiparticle then meets with a particle that makes up part of the black hole. These two particles then annihilate each other and—poof!—there goes a little mass of the black hole. That is, black holes lose mass—they evaporate.[2]

Not-So-Black Holes

What this means is that black holes aren't truly black, and through these quantum fluctuations and escaping particles, they glow. And as they glow, they steadily lose mass. This Hawking radiation is a feeble thing, with black holes emitting only a tiny morsel of radiation. A black hole with the mass of the Sun will take more than 10^{60} years to reduce its mass by 1 percent.

But Hawking radiation has a strange property in that the amount of radiation is inversely proportional to the mass of the black hole. This is mathematics speak for "the smaller the black hole, the more intense the amount of Hawking radiation."

So here we have a feedback loop. Due to Hawking radiation, a black hole will lose mass over time. And as it loses mass, its Hawking radiation increases, so it loses more mass, so its Hawking radiation increases. This runaway process continues, with the glow of Hawking radiation increasing until the black hole actually becomes visible. By this stage, the mass of the black hole is decreasing rapidly, and the intense radiation becomes ultraviolet, then X-rays and then gamma rays, and then pop! The black hole has completely evaporated, and nothing is left.

How long do we have to wait for a black hole to evaporate

away to nothing? Much, much longer than the time it takes for all the protons to decay. For a black hole with the mass of the Sun, it is about 10^{67} years, meaning that long after the last star has been extinguished and long after the last matter has melted away into the background, every so often in the utterly dark universe, there will be a short, sharp flash of light as a black hole decays away.

Supermassive black holes, with masses a billion times the mass of the Sun, are known to exist at the centers of the most massive galaxies, and it will take even more time for these to evaporate away their mass. These supermassive black holes might last for up to 10^{100} years, but their time will come, and they, too, will eventually vanish from the universe in a puff of Hawking radiation.

Once the last black hole has decayed away and the final fireworks of Hawking radiation are lost forever, the darkness of a never-ending night will fall. It appears to be the end of all things.

Or is it?

Is the end of the universe really the end?

Once all the dead stars have dissolved and the black holes have evaporated, there will be no more stuff in the universe.[1] Well, no substantial stuff. All that will exist is an ever-cooling, ever-thinning soup of electrons and photons. The universe will be cold and uniform, and all useful energy will be gone. Without this energy, there will be no prospects for complexity and life. The universe will have reached its ultimate state, its *heat death*.[2]

The heat death sounds dramatic, but it was predicted long before modern cosmology with its fiery birth and expanding spacetime. In 1851, Lord Kelvin (whom we briefly met earlier) proposed that the universe was running down, cooling from hot to cold, with the eventual fate of heat death ahead of it. He was not the first to propose this idea, but he was the first to set it in the language of thermodynamics.

We already considered the second law of thermodynamics when we were wondering why the universe was so similar everywhere and how it means that energy in a process is eventually minimized. We need to understand this in a little more detail, and to do so, we will have to introduce another concept from thermodynamics, namely *entropy*. An awful lot of words have been written about entropy and what it means. Words like *disorder* are thrown about, exemplified by the messiness of a teenager's bedroom. The concept of entropy can indeed seem messy and confused. This is not helped by the fact that there is not just one definition of entropy.

Ludwig Boltzmann wrote down the first mathematical definition of entropy in the 1870s, an equation that now adorns his gravestone. At the same time, a slightly different formulation of entropy, based upon probabilities, was written down by the American statistician J. Willard Gibbs. The two sets of mathematics look very similar, but there are occasions when the answer you get depends upon the mathematics you choose to use. Obviously, this is not ideal.

Both Boltzmann and Gibbs were interested in thermodynamics, the flow of heat in a process, which was born out of a truly practical problem during the Industrial Revolution, namely determining how efficient a steam engine can be. It then grew into an edifice of modern science that has occupied the minds of top physicists ever since. A student in high school might first encounter thermodynamics in the form of the gas laws of Boyle, but once they get their teeth into the subject at college, it becomes statistical physics, examining the different ways that atoms in a gas might be rearranged to produce similar or different outcomes. At its heart, though, thermodynamics is a study of the flow of heat.

A neat way to think of the entropy of Boltzmann and Gibbs is to consider the amount of useful energy in a system. "Useful" here means that the energy can be tapped to do something, such as run an engine or power a living being. Things that have lots of useful energy are at lower entropy than those at high entropy. Notice that it's not the *total* amount of energy that is important, just the amount of *useful* energy.

You might be scratching your head over this a little, so let's think of a simple example. Imagine you have two blocks of metal, one hot and one cold. If you connect the two, heat will flow from the hot block to the cold block, and you could theoretically use this flow of energy to power an engine. So the situation with a hot block and a cold block of metal is at low entropy as there is energy that can be used. If, however, we have two warm blocks, containing the same total amount of energy as the hot block and the cold block, when we connect them, no heat flows between them and there is no useful energy in the system. The two warm blocks are at higher entropy. This might seem a little esoteric, but in fact, we can think of any process in terms of its flow of energy from one place to the other.

Humans experience the flow of heat, the energy that powers

our everyday world, as an irreversible process. Shown a video of a food cooking, an egg cracking, or a vase breaking in reverse, we immediately feel a sense of dissonance. But show someone a video of a ball arcing through the air, and they will not be able to tell the difference between forward and reverse. All our tried-and-true laws of physics are *reversible*. Play the laws of physics in reverse, and they are still solutions to the equations. The laws of physics are time *symmetric*. But how can it be, then, if—according to the laws of physics—going forward is just as good a solution as going backward, time flows in only one direction? This paradox is probably the most obvious and simplest to state unsolved problem in physics.

The so-called *arrow of time* gives a memorable name to the idea of the asymmetry of time, that it has a definite unidirectional flow.[3] This idea is at odds with the equations of motion of Newton, Maxwell, Einstein, and even Schrödinger, all of which are time symmetric. There is one irreversible process in quantum physics that we haven't mentioned yet, though, and that is observation.

Collapsing the Wave Function

We know from our discussion of Heisenberg's uncertainty argument that measuring one property of a system can unavoidably affect another. Heisenberg's example was position and velocity. A more extreme example is measuring the position of a photon by absorbing it—and hence completely destroying it! This doesn't sound all that reversible, does it? In fact, the technical name for the mathematical operation of measurement is *collapsing* the wave function.

Today, when an unwitting undergraduate physics student is introduced to quantum physics, they are presented with the *postulates*. These are the rules of quantum physics stated in a concise and bite-size way, far removed from the tumultuous path that actually brought us here. The three main postulates are as follows:

1. When a physical system is prepared, it is mathematically described by its quantum state (i.e., the wave function).
2. As time passes, the quantum state changes according to the Schrödinger equation.
3. When a measurement is made, the quantum state immediately becomes the one associated with the outcome observed.

Really, the first and the third are essentially the same if you imagine filtering out the outcome that you want to prepare the system in. Still, we are left with two different ways things can happen to a quantum state.

The Schrödinger equation we met earlier is the bedrock of quantum physics. It is time symmetric and unambiguously applied. The collapse is another beast altogether and the cause of great strife in the foundations of quantum physics. For the practitioner, it is obvious when to apply it—when a measurement is made. However, for a third party, it is not so clear. When exactly is the measurement made, and who qualifies to make one? Must it be a scientist? Does nature itself measure quantum systems? All these questions encompass the so-called *measurement problem*. Since all the successful equations of motion in physics are reversible, physicists generally do not like the fact that the seemingly violent collapse of the wave function is irreversible. It's a problem with the theory, they say.

Why is it still there then? Well, it works—and does so with astonishing precision. But there also is no consensus on which process is more fundamental, the Schrödinger equation or the measurement. If we accept that collapse is a real part of the physics and not some artifact we haven't figured out how to get rid of, then we have found a source of irreversibility! But not so fast. After all, you—the *thing* doing the measuring—are made of atoms and thus must be described by quantum physics as well. Indeed, to measure something is to interact with it, and interactions are again described by the Schrödinger equation. We've come full circle!

The task of twentieth-century quantum physicists was to either come up with the precise location in which the quantum state collapsed or show that the reversible dynamics of the Schrödinger equation can lead to increasing entropy. More than one hundred years later, we still cannot say when and where the quantum state collapses. If we take an operational view of quantum physics—a view that presents the theory only as a toolbox for practitioners—then we can trace the collapse to some internal update happening in the mind of the observer, or an observer observing an observer, or...well, you get the picture. Luckily, the second path has provided some fruitful answers in a disguise we have seen before: entanglement.

When two systems interact, they generally become entangled. The nature of entanglement means all the information is carried by the entire system, and—in the extreme cases—the individual systems contain zero information. As systems continue to bump into one another, entanglement builds and builds until we reach the point that *any* particular part of the system possesses zero information. To connect this back to thermodynamics, consider that one would certainly need information about a system in order to extract useful energy from it. While a more formal connection between thermodynamic quantities like energy and heat can be made, the crux is just that—no information, no useful energy, maximum entropy.

Now, that quantum system could very well be the universe as a whole. In other words, the universe—taken to be one big quantum thing composed of smaller quantum things—ends up a big entangled mess. Even in the quantum realm, entropy increase marches on, pointing steadfastly in the direction of the arrow of time.

No matter how many times physicists have tried to reverse the arrow of time and explore the origin of our universe (theoretically, anyway), we keep coming back to Kelvin—the fact that everything is running down goes right through the heart of physics. Our universe was born with lots of useful energy: smoothly distributed matter with the potential to collapse into stars and light elements with the potential to undergo nuclear fusion into heavier elements. *Why* our universe was born with its abundance of useful energy is a mystery, but every moment of every day, this useful energy is dwindling. Even you reading this page is processing lower entropy energy, probably from that yummy burger you had for lunch, into higher entropy, less useful energy, as the infrared radiation emitted from your glowing skin.

This process continues forever and is irreversible. Once the last proton has decayed and the last black hole has evaporated, Kelvin's nightmare of a cold, dead universe will be inevitable. Perhaps the universe will have entered its final state, and that's all there will ever be.

But perhaps not! The universe might well be reborn. And the secret ingredients to allow it to do so might be dark energy and the quantum.

Darkness as the Antidote to Darkness

We've met dark energy in passing a few times in this book. Remember, it's an energy density that fills all of space, but it is unlike any other energy we know. Its peculiar properties mean that dark energy is driving the cosmic expansion faster and faster, accelerating the demise into the heat death as it thins out matter and radiation at an ever-increasing rate.

A peculiarity of dark energy is how it behaves as the universe expands. If the universe doubles in size, the density of matter drops by a factor of eight due to the growth in volume. Radiation, such as light, also thins out as the universe expands, but faster than matter does. But dark energy doesn't dilute at all.

Today, the density of dark energy is equivalent to about 10^{-29} g/cm^3, which, while small, is about twice the average density of matter in the universe. And dark energy will have the same energy density in the future, even the dim and distant future of the heat death of the universe. So lurking in the background of the apparently dead universe will be dark energy. *Interesting,* you might be thinking, *but so what?*

Remember at the start of this book, when we examined the quantum at the start of the universe, we presented the inflaton, an energy field that drove super-rapid expansion before the first matter and radiation existed in the universe. One possibility for the mechanism behind inflation is that it is an energy field that underwent a change due to the action of quantum tunneling, moving from one state to another through a process that is impossible without the actions of quantum mechanics.

Some physicists speculate that dark energy might be in a similar situation, existing in an energy state that is not its true minimal value, but it is stuck there, in something known as the *false vacuum,* as there is no process that will allow it to decay. But remember, quantum mechanics does offer the opportunity for things to undertake impossible transitions through quantum tunneling. Perhaps dark energy can undergo a transition from a higher state to a lower state?

Like all of quantum mechanics, tunneling is a probabilistic thing, and the chance of a dark energy transition increases the longer we wait. How long will we have to wait for dark energy to undergo such a quantum decay? Well, we're well within the bounds of speculation, and any estimate of how long you have to wait has to be taken with a large pinch of salt, but some have suggested that on a timescale approaching $10^{1,500}$ years, the dark energy false vacuum will ultimately decay into a true empty vacuum.

Just how this decay will proceed is even more speculative, but there are some interesting ideas. One is that the decay will not occur simultaneously across the entire universe but that different patches of the universe will undergo decays at different times. This is how water freezes as it is chilled, with freezing starting at separate points and spreading outward until the entire body of water is frozen.

As the dark energy decays away, it may act like the inflaton we saw in the earlier chapters, driving a bout of accelerated cosmic expansion, and essentially, at each location in our existing universe, a new universe is born.[4] What will these new universes be like? Again, speculation must be heaped upon speculation, but all these baby universes could be just like ours. A more interesting idea, however, is that the act of crystalizing a universe out of this new inflation actually rewrites the laws of physics, with each universe having its own unique blend of particles and forces. Most of the universes could be very unlike our own, being probably too simple for the complexity required for life, but in some of them, stars could shine, planets could revolve, and life of some sort could form. Some life-forms might even learn to read and write!

There are a few scientists who even think this has already happened! Maybe our universe is just one stage of an eternal cycle of universes being born, living, and then dying and by doing so giving birth to new universes. Or maybe not. For now, the far future of our universe and what will become of it are little more than fairy tales.

PART 4

The FUTURE of a QUANTUM COSMOS

Where have we been?

Through this book, we have ranged over the past, present, and future history of the universe, at least as we understand it. We have seen that as much as quantum physics and general relativity are different, they are inseparable if we are to understand the inner workings of our cosmos today.

From a birth in the Big Bang through our present epoch where stars shine brightly and life thrives (at least on this one little planet), we can unravel the inner workings of the cosmos. Even into the distant future universe, we can guess at the dark times ahead to a time with no shining stars and eventually to when matter melts into the darkness. We can see into the past and future using the laws of physics, charting the times no human has experienced through the language of mathematics. We are time travelers through the imaginings of astronomers and cosmologists.

The fact that we can have any confidence about the state of the universe a fraction of a second after the Big Bang or even many trillions of years into a future none of us will ever experience demonstrates the success of modern science. This is the same science that underwrites the technology of modern civilization, that provides machines that can see inside an ailing body, that delivers almost unlimited information into a tiny computer that you can hold in the palm of your hand, and that can ensure that you never ever get lost again on the way to a party. Modern science is something all humans should celebrate.

But science is never complete, and there are always questions left to answer. And our understanding of the fundamental universe is far from done. At the start of this book, we pointed out that modern physics is built on two seemingly incompatible ideas. Gravity, which dominates the large-scale universe, is written in the mathematics of Einstein's general theory of relativity, whereas the other forces—electromagnetism and the weak and strong nuclear forces—are couched in the language of quantum mechanics.

It should be clear from the preceding chapters that in describing the past, present, and future universe, cosmologists must somehow glue these two seemingly disparate ideas together. In many cases, they can get away with it, since while both gravity and the other forces can be very important, they can often be treated independently. But in other cases, they are so intertwined that the quantum influences gravity, and gravity influences the quantum. It is in these places—the hearts of black holes and the birth of the universe—that the mysteries of the universe remain.

In the remaining few pages of this book, we'll try and look into a more immediate future—our scientific future—and think about what the next insight might be and what that could reveal to us about the mysteries of the universe. But let's start with a dream.

Physicists have a dream, a dream of a theory of everything. What they hope to discover is a single set of mathematics that describes the influence of gravity and the quantum forces wrapped up into one set of equations. The hope is that with this one set of mathematics, all the mysteries of the universe will be revealed. We will see what happens in the most mysterious places, including the centers of black holes, and we will truly understand where our universe came from.

This quest for a theory of everything has occupied many physicists for many years. Even Einstein tried to weave the forces of electricity and magnetism into his description of curved and warped space and time. Others tried too. Almost as soon as relativity was discovered, Theodor Kaluza and Oscar Klein, working separately, attempted to write electromagnetism and quantum mechanics into the form of extra dimensions additional to Einstein's four-dimensional space and time.

Einstein's quest continues to this day as scientists attempt to unify all the fundamental forces. Physicists have tried different approaches and considered various assumptions in their quest to build new mathematics for the universe. Some have pushed on the mathematics we know, adding new pieces to see if that can give new insights. Others have tried to chop the particles of the universe into new, smaller pieces and build new physics from the ground up. Others have gone further and sliced and diced space-time itself into discrete little chunks so that the universe itself emerges from a more fundamental construction. But so far, all attempts have failed. Maybe someone reading these lines will finally make the crucial breakthrough! Just what is needed?

What does a theory of everything look like?

Let's take a whirlwind tour of some of the ideas that physicists have explored in their search for a theory of everything.[1] This is not meant to be a comprehensive list but a taster, a summary of the concepts that have made it into the public consciousness. However, it is important to remember that they are not all independent, and the mathematics can be interlinked and intertwined as ideas flow back and forth.

Supersymmetry

We've already mentioned how much physicists love symmetry. Symmetry leads to beautiful equations, conservation laws, and a simpler picture of the universe. There appear to be symmetries underlying the standard model of particle physics, with six types of quarks squared off with six types of leptons, each arranged in pairs of increasing mass, known as families. Physicists have learned to write about the properties of these in terms of the language of *group theory*, which encompasses such symmetries.

Some physicists have wondered if we can expand the standard model by imposing more symmetry on it with the additional particles that result. With this, the electron has a supersymmetric counterpart, the *selectron*, and for each quark, there is a *squark*. Other particles have supersymmetric fellows: Ws and Zs have *winos* and *zinos*, and perhaps, somewhere in this mix, is the *graviton*, the particle that carries the force of gravity.

The theory appears to be mathematically elegant, tying up many loose ends. Unfortunately, it appears to be completely wrong. There is no evidence for the existence of selectrons and squarks, with none ever having been spat out of the Large Hadron Collider, the largest scientific experiment ever created, which was constructed to test the limits of the standard model. Desperate physicists have suggested that the supersymmetric particles are really massive and difficult to produce at the energies of the Large Hadron Collider at CERN, but to make this supposition, some of the underlying symmetries have to be broken. For a theory that is supposed to be supersymmetric, this is a big break in logic. While there are still people plugging away at the mathematics, trying to make supersymmetry work, many have concluded that it is not the path into the light.

String Theory

String theory attempts to unify gravity and the other forces by going down to the truly fundamental level. In this picture, at the smallest level, everything is made of the same stuff, tiny vibrating strings. So electrons are made of strings, quarks are made of strings, and the thing that tells you how an object is defined is the vibration of the string. Now, this might seem a little crazy, but there are some mathematical properties of these vibrating strings that are very enticing to physicists, making them look a lot like the particles we see around us.

One of the things that is built in to string theory is the force of gravity, in which the graviton can be one of these vibrating strings. This picture sounds oh so simple, with everything—absolutely everything—made of exactly the same stuff at the bottom level. But the mathematics needed to make string theory function is messy. One of the messiest features is that you need to add extra dimensions for the strings to vibrate in. Not just one or two or three but, in some versions of string theory, your universe needs a total of maybe twenty-six dimensions.

"Where are these dimensions?" detractors cry. But those working with strings have added a fudge called *compactification* where any unwanted dimension—that is, one we don't experience in everyday life—is neatly rolled up not to bother you. String theorists march on with the mathematics, but it turns out the expected size of the strings is so small there is no hope of the Large Hadron Collider seeing them. In fact, if you wanted to test the experimental impact of strings, you would need a collider the size of our Milky Way galaxy, and that is somewhat impractical. String theorists are left playing in the math. Without an experimental test to provide supporting evidence, the string theory naysayers declare that string theory isn't even real science!

M-Branes

The failure of string theory to produce a theory of everything doesn't mean people have stopped pushing the boundaries. String theory has grown into M-theory, introduced and named in 1995 by Edward Witten, who suggested the *M* should stand for magic, mystery, or membrane, according to the taste of the reader. Coy attempts at humor aside, the underlying idea is that what we think of as strings, which are one-dimensional objects, are really drawn out into extended structures—membranes (or branes for short)— that are floating around and interacting in some higher dimensional space. Like strings, these branes make up everything and are the fundamental pieces of the universe.

Like string theory, the hope is that somewhere in the mathematics there is a particular form of vibrating membrane that will account for the action of gravity. But again, the mathematics required is fiendish, and there is not really a single idea that makes up M-theory. There are a whole host of them, built upon differing ideas, each with its own set of assumptions and restrictions.

It is not only the difficult mathematics that M-theory shares with string theory but also the limitations on testing whether any of the mathematics that are now poured over thousands of pages of academic journals have anything at all to do with reality and the physical world around us. Particle accelerators and gravitational wave detectors have provided no evidence to support M-theory, while in offices and on whiteboards around the world, some of the smartest minds continue to bend and stretch and push the theoretical mathematics. Maybe one day M-theory will turn out to be our ultimate description of reality, or maybe it will eventually fade away as people become bored and disheartened with the lack of prediction and experimental evidence.

Loop Quantum Gravity

Some physicists have wondered if there was another way to unite the gravity with the quantized nature of the other forces. Perhaps, some reasoned, gravity is a quantum phenomenon but not in the way that the other forces are quantum phenomena. Einstein told us that gravity was the result of curved space and time, so what if we quantize space and time themselves, chopping them into little chunks? This is the start of *loop quantum gravity*.

As it contains curved but dissected space and time, gravity is already present, while the other forces play out on the quantized backdrop. You might be asking where the "loop" comes from in the name of this theory, but be warned, this is where this idea becomes a little weird. The little chunks of space and time are effectively woven together in a matrix or network, so if we could look deeply into the smallest bits of space and time, they would look like a close-up of a woolly sweater, with loops wrapped through loops holding everything together. There are even ideas on how the future grows out of this network, like a loom adding a new line of woven thread to material.

One of the philosophical issues that great minds have battled with since Einstein wrote down his equations is that the past, present, and future are all there, writ large in the mathematics. There is no unfolding future in this *block universe* of general relativity, and as the future is already written, the question of our free will and our experience of time is questioned. Any theory that is written onto this space-time fabric will also face these questions. "Not us!" the loop quantum gravity people tell us, because for them, the future has yet to be woven, and free will is safe! But like string theory and M-theory, the mathematics is hard, the ideas are incomplete, and the experimental evidence for loop quantum gravity remains missing in action.

And Many Other Theories

We've covered a lot of ground here but have not considered all the possible routes to a theory of everything. In fact, physicists are quite desperate in their search for a solution, and many different ideas, based upon quite radical views of the fundamental makeup of the universe, are out there. Most get way more press coverage than their true predictive power or relation to reality should really give them credit for, as a reader of the casual scientific media might conclude. The truth is that we have been steering somewhat blind for more than half a century, and while the number of words and equations written might be continually increasing, we appear to be no closer to the solution.

Where can a theory of everything take us?

After a century, all the great human minds have failed to provide a convincing theory of everything. But we can still dream! What if we wake up tomorrow and discover that someone has cracked the puzzle, and we now have a single theory that encompasses gravity and the other forces? It is impossible to guess when or where this solution will come from, but some young researcher in a physics and mathematics department somewhere in the world will have a truly light-bulb moment where everything becomes crystal clear. That person will be on the road to a Nobel Prize. But what will we have learned? What could we discover about the universe?

We've already mentioned two places: the centers of black holes and the birth of the universe. What could we find there?

In Einstein's general theory of relativity, once mass collapses below a critical point so that it is all inside the event horizon, there is nothing to prevent the continued collapse down to nothing, to the formation of the singularity. With only Einstein's relativity on the table, it would seem that gravity can always overwhelm the other forces, squeezing and squeezing mass into a zero-sized volume. But with our theory of everything, we will understand the true relationship between the forces. Many physicists feel that when the densities of matter get extraordinarily high, as we would see in the formation of a black hole, it would not be simply a case of gravity dominating and the other forces becoming irrelevant. Instead, as gravity grows, so do the other forces, fighting the collapse to nothing. Instead of forming an infinitely dense singularity at the heart of the black hole, the action of the quantum forces could produce a supersmall, superdense core, with the infinities from Einstein's gravity banished. While bizarre and extreme, the black hole will not have infinities. Without having to worry about there being real, physical infinities in the universe, physicists will be able to sleep well in their beds.

Do we gain any further understanding about black holes from a potential theory of everything? At this point, we definitely move into the realm of speculation, and the extremely dense but not quite infinite black hole core might be all we have. But some think that other, stranger things might be possible. The extreme gravity at the center of the black hole might punch a hole in the fabric of the universe, creating a *wormhole* to another location or another time or even another universe. While this sounds like the stuff of science fiction, Einstein's mathematics hints at the existence of such weird possibilities, and fiction might one day become reality.

What about the birth of the universe? According to Einstein, all space, all time, all matter came into existence at the Big Bang, but as with the centers of black holes, our mathematics is faced with infinities as gravity dominates and the other forces take a back seat. As with black holes, we expect these infinities to be banished once the true relationships between gravity and the other forces are understood. What do we expect this to reveal?

Perhaps the rough picture of Einstein is correct. Perhaps space and time and matter all came into being at the initial start time of the universe. Perhaps the mathematics can go no further, and there is no answer to the question "where did the universe come from?" Most physicists find this idea unpalatable and don't think that is likely to be the case.

Looking at the hints in Einstein's mathematics, many think our universe was not the actual beginning of everything and that we come from some preexisting structure. As we touched on in the last chapter, the ultimate demise of our universe might lead to the birth of new universes, and this process might be where our universe came from. Or perhaps it was born in the death of a massive star in a preexisting universe, a star whose collapse formed a new black hole and, in the act, budded off a new universe.

Or perhaps our universe was born out of a process we can barely imagine. At the moment, without the mathematical language of gravity and the other forces working together, all we can do is guess.

What else can we expect from a theory of everything? Remember, we have several holes in our understanding of the universe, holes that this new theory is expected to plug. Specifically, we cannot currently account for the dark side of the universe, the dark energy and matter that dominate the large-scale cosmos, within our standard model of particle physics. The standard model is extremely successful at explaining all the things spat out by the Large Hadron Collider, but there is no explanation for any of the dark stuff.

As we have seen, many physicists have suggested extensions to the standard model, but as yet, none have predicted a dark matter particle that has been observed in any accelerator or singled out in any astronomical observation.

Dark energy represents an even bigger problem, as the universe would have happily existed without it. Why does it exist at all? At the moment, many think it is something to do with the quantum nature of the vacuum, but all our theoretical calculations are woefully inadequate at making any kind of predictions about its nature. Perhaps, with a theory of everything in hand, all the pieces will fall into place, and we'll see that dark energy is just a natural bit of our universe with a specific purpose.

What still stands in our way?

You might be wondering why it has been so difficult to create the theory of everything. Why can't some smart physicists simply think really hard and come up with a nice shiny theory, something that will encompass all the forces, the dark side of the universe, and much, much more?

The problem is that at the moment, physics is faced with a rather embarrassing issue. We've already noted that modern physics is built upon general relativity and quantum mechanics, and within their own domains, each discipline is extremely successful. What this means is that every experimental test we throw at them, they pass with flying colors.

In the last decade, the discovery of the Higgs boson at the Large Hadron Collider represented the cherry on the top for the standard model of particle physics. Every time physicists power up a particle accelerator, what comes out aligns with the mathematical predictions of the standard model.

For general relativity, the picture is the same. Its cherry moment was the discovery of gravitational waves in 2016. These tiny ripples in space and time themselves are born in some of the most violent and energetic events in the universe. But due to the weakness of gravity, they carry only feeble amounts of energy across the universe. And after more than half a century of hard work, false starts, and mistaken claims, the Laser Interferometer Gravitational-Wave Observatory (LIGO) registered the signature of two merging black holes in the distant universe. For this discovery, some of the pioneers behind LIGO—Rainer Weiss, Kip Thorne, and Barry C. Barish—were awarded the 2017 Nobel Prize in Physics.

In the last few years, LIGO has become just another astronomical observatory, scanning the skies and picking up the signals of energetic events across the universe. People barely bat an eyelid now at the newest discovery as detections become routine. But as the data come in, the precise signal is pored over to see if there are any deviations from the predictions of Einstein's great work. And while there are hints and possibilities, always down at the detectability limit of the instruments, Einstein's general relativity appears to win every single time.

These words are being written in 2020, a year that is likely to be remembered for a long time, and not because of quantum mechanics. In the months before the COVID-19 pandemic struck, astronomers reported their new observations of a binary pulsar system. This system is two superdense stars, the leftover remnants of a massive star that died, that individually spin very rapidly as they orbit each other at extremely high speed. Here you have all the ingredients for relativity, including rapid speeds and spins and massive gravity due to dense stars. To understand and predict the stars' motions, physicists need to consider the immense bending and warping of space and time. They even need to understand an esoteric phenomenon known as *frame dragging*, where space and time are pulled around by the orbiting stars. This causes the orientation of the spin of each star to steadily change over time in a way that is different from the celestial mechanics of Newton. And just what did astronomers measure? You guessed it—that Einstein's predictions were correct again. You would think that physicists would be ecstatically happy about this new development, but as we've pointed out, this is actually a very unsatisfactory situation, as we know both quantum mechanics and general relativity on their own cannot function as a comprehensive explanation of the universe. There must be something else. But nature is not giving us the clues to take the next step.

Our most successful physical theories were born out of necessity to explain unaccounted-for *observations*. What physicists yearn for is guidance on what to do next. What they want is a clue to where these theories cannot account for nature. In this book, we've explored such places, such as at the centers of black holes and the birth of the universe, but the infinities we have found there so far really don't help. What physicists really want is something they can work with.

What kind of thing, you may ask? Something like a new, unexplained particle being produced in a collision at the Large Hadron Collider. Or a gravitational wave signature that cannot be explained by Einstein's description of two merging massive objects. Physicists are desperate for anomalies, the unexpected, the unexplained.

There are hints of odd things all the time, but they usually don't go anywhere. Something strange appears in the noise of an experiment or observation. Maybe there's an unusual-looking wiggle in the gravitational wave signature of a merging pair of black holes. Perhaps there's an unexpected bump in the energy distribution of photons coming from a particular interaction at the Large Hadron Collider. Upon the discovery of such anomalies, theoretical physicists kick into overdrive, squeezing and molding their favorite theoretical ideas that are "beyond the standard model" to see if this is evidence that they might be right. The activity can be frenetic and impressive—try googling "diphoton excess" to see an example of this in action. But as more data comes in and the inevitable noise of experiment and observations is beaten down, these strange anomalous signals usually vanish, a remnant of a statistical fluke in the data. And as they do, the excited cries of "new physics" tend to vanish with them.

What should physicists do next? Honestly, they are not too

sure. For some, the answers will come with bigger telescopes to scan the heavens or more powerful particle accelerators to reveal the inner workings of the smallest particles. But there is the constant worry that even with these new eyes on the universe, relativity will continue to explain the cosmos, and the quantum will explain the quark, with no unifying theory bringing them together.

In fact, our telescopes and accelerators may never be powerful enough to reveal the theory of everything. Some think that when it comes to scientific instruments, bigger is not really better and that we need to *think* harder—that new breakthroughs in mathematics and logic will provide new tests that are possible in a desktop laboratory. But this, too, is more of a dream than a plan.

So this is where we find ourselves. Our physical laws are dominated by two incompatible theories, one describing the big, the other the small. Both of these theories work surprisingly well when kept to their domains and, as explored in this book, can be jerry-rigged together where needed. But ultimately, both must be incomplete.

Physicists are a stoic lot, and we think it is important to finish this book on a positive note. The last few centuries have seen a revolution in our understanding of the universe, from the very small to the very large. This is an incredible achievement of which we should all be proud. But the journey is not over, and there are questions left to answer, including that of the true relationship between the quark and the cosmos.

Every day, around the world, smart young people are being drawn into physics, some after hearing about this conflict between the forces of the universe. A breakthrough could come at any time, with a new observation or experimental result or a new development in the theoretical predictions. For now, whether we can truly link the quantum and the cosmos is a waiting game. As we wait for the right thought or unexplained observation, one thing is for sure—when it comes, it will completely change our view of the universe. Besides allowing us to peer into the mysterious hearts of black holes and into the very birth of the cosmos or revealing the nature of dark matter and energy, it is likely to tell us so much more. Perhaps it will even take us closer to the big questions about how life begins, whether we are alone in this universe, and why we exist to dream up questions that touch upon every corner of the cosmos and beyond.

Acknowledgments

Chris thanks his colleagues, friends, and family for their constant encouragement. Lindsay, you have been my greatest supporter and gentle critic. Dylan, Max, Wes, and Evan, you are my inspiration. I hope *Quantum Physics for Babies* has prepared you well for this! Thanks to Geraint for immediately signing on to this (what I thought was a) crazy idea. Thank you to the whole Sourcebooks team, who have been amazing partners over the years.

Geraint would like to thank Chris for inviting him to collaborate on this book, to bring together the world of the very large and the very small. It has been an immensely enjoyable journey. Geraint would also like to thank Luke Barnes, Jon Sharp, Sally Bennett, and Matt and Jo Wilken for their comments, insight, and support in bringing this book to completion. The love of my family—Zdenka, Bryn, and Dylan—has been invaluable, even with us all squeezed into our small apartment, distance learning and distance working. This year, 2020, is one that will not easily be forgotten, and the crazy times seem yet to continue. But, friends, when it is over, we will meet again.

We are deeply indebted to Anna Michels and the editorial team at Sourcebooks for taking the esoteric ramblings of two physicists and turning them into something palatable.

Endnotes

There are many sources of information on the physics of the quantum and the cosmos. Below we include some key sources of extra reading (or watching) on the topics covered in this book.

The Quantum and the Cosmos

1   *Lord Kelvin's prediction of the end of physics*: The history of science is always messy, and no one is sure if Kelvin actually said his famous quote on the end of science, but he was an intriguing scientist. See David Saxon, "In Praise of Lord Kelvin," *Physics World*, December 17, 2007, https://physicsworld.com/a/in-praise-of-lord-kelvin.

2   *The conservative Planck*: This is now legend and succinctly described by his colleague and friend (and fellow quantum founder) Max Born in "Max Karl Ernst Ludwig Planck,

1858–1947," *Obituary Notices of Fellows of the Royal Society* 6, no. 17 (November 1948): 161–88, https://doi.org/10.1098/rsbm.1948.0024.

3   *On the most famous scientist to have lived*: Abraham Pais's *Subtle Is the Lord: The Science and the Life of Albert Einstein* (London: Oxford University Press, 2005) charts the life and career of Albert Einstein, including his contributions to physics in 1905, his "miraculous year," and his rise to fame after the experimental verification of his general theory of relativity.

4   *More on Maxwell*: Robyn Arianrhod's *Einstein's Heroes: Imagining the World through the Language of Mathematics* (Brisbane: University of Queensland Press, 2003) is a biography of James Clerk Maxwell, the so-called father of modern electromagnetism. It is from his insights that Einstein's revolutionary view of the nature of light was born.

5   *More on Einstein's theories*: For a take on relativity in his own words, see Albert Einstein, *Relativity: The Special and the General Theory* (many editions, including Princeton University Press's 100th Anniversary edition, 2019).

6   *Our view of modern cosmology*: There are many discussions on the development of our view of the universe. "Cosmology and the Origin of the Universe: Historical and Conceptual Perspectives" by Helge Kragh (arXiv preprint, 2017, https://arxiv.org/abs/1706.00726) is an excellent introduction, while *The Cosmic Revolutionary's Handbook (Or: How to Beat the Big*

Bang) by Luke A. Barnes and Geraint F. Lewis (Cambridge: Cambridge University Press, 2020) outlines how scientists use the observations of the universe to reveal its inner workings.

PART 1: THE QUANTUM OF COSMOS PAST

Where did the universe come from?

1   *How do we know there are two trillion galaxies in the universe*: For a popular science–level discussion, see Ethan Siegel, "This Is How We Know There Are Two Trillion Galaxies in the Universe," *Forbes*, October 18, 2018, https://www.forbes.com /sites/startswithabang/2018/10/18/this-is-how-we-know-there -are-two-trillion-galaxies-in-the-universe.

2   *Who discovered the expanding universe*: Again, the history of science remains messy, and the question of who should get credit for discovering the expansion of the universe is not a simple one. Harry Nussbaumer and Lydia Bieri's "Who Discovered the Expanding Universe?" (arXiv preprint, January 16, 2012, https://arxiv.org/abs/1107.2281v3) unravels the history.

3   *Quantum fluctuations:* The idea of "vacuum fluctuations" can be traced back to English physicist Paul Dirac, though it wasn't popularized until later. See P. A. M. Dirac to Niels Bohr, August 10, 1933, Niels Bohr Library, American Institute of Physics, College Park, Maryland.

4   *The never-ceasing movement of quantum energy:* The existence

of quantum fluctuations was popularized by Stephen Hawking in "Breakdown of Predictability in Gravitational Collapse," *Physical Review D* 14, no. 10 (November 1976): 2460–73, https://doi.org/10.1103/PhysRevD.14.2460.

5   *On the birth of matrix mechanics*: Now considered legend, Heisenberg later recalled his epiphany arising from a late-night calculation on the island of Helgoland. See Werner Heisenberg, *Der Teil und das Ganze* [Physics and Beyond] (Munich: Piper, 1969).

6   *Quantum history*: A detailed discussion of the development of quantum mechanics, including Heisenberg's revelation, can be found in Manjit Kumar's *Quantum: Einstein, Bohr, and the Great Debate about the Nature of Reality* (London: Icon Books, 2008).

7   *A universe from nothing*: See Edward P. Tryon, "Is the Universe a Vacuum Fluctuation?," *Nature* 246 (1973), 396–97, https://doi.org/10.1038/246396a0.

8   *Most cosmologists are not satisfied*: Despite Lawrence Krauss's popular book *A Universe from Nothing* (New York: Atria, 2012), there are numerous alternatives proposed for what came before the Big Bang, from our universe being just one in an infinite sequence of repeating universes (such as in ekpy-rotic cosmology) to cosmological natural selection, where universes give birth to other universes through black holes, to the concept of the multiverse and our being one of many universes

out there. We simply don't know. See a recent conversation on the topic here: Stephanie Pappas, "What Happened before the Big Bang?," Live Science, April 17, 2019, https://www.livescience.com/65254-what-happened-before-big-big.html.

9   *The multiverse*: This is not a single concept in physics but more of a grab bag of ideas. Professor Max Tegmark has a nice review (you have to excuse the outdated website design): Max Tegmark, "Welcome to My Crazy Universe," Universe of Max Tegmark, accessed November 23, 2020, https://space.mit.edu/home/tegmark/crazy.html.

Why is the universe so smooth?

1   *In the beginning*: A detailed but gentle walkthrough of the initial state of the universe can be found in Steven Weinberg, *The First Three Minutes: A Modern View of the Origin of the Universe*, 2nd ed. (New York: Basic Books, 1993). And if you want the gory details, see Edward W. Kolb and Michael S. Turner, *The Early Universe* (New York: CRC Press, 1994).

2   *The standard model*: The Conseil Européen pour la Recherche Nucléaire (CERN) is an international collaboration of scientists that has produced many of the most important discoveries in science. It also maintains a great deal of educational material on high-energy physics. For more on the standard model, see "The Standard Model," CERN, accessed November 23, 2020, https://home.cern/science/physics/standard-model.

3   *Higgs boson*: There is no standard for how things are named in science, but sometimes they are named after one of the scientists involved in the discovery. The physics here is generally credited to six scientists: Robert Brout, François Englert, Peter Higgs, Gerald Guralnik, C. Richard Hagen, and Tom Kibble. Only two won the 2013 Nobel Prize in Physics. For a discussion of the politics of science, see Joel Achenbach, "Nobel Committee's 'Rule of Three' Means Some Higgs Boson Scientists Were Left Out," *Washington Post*, October 8, 2013, https://www.washingtonpost.com/national /health-science/peter-higgs-francois-englert-win-nobel-prize -in-physics/2013/10/08/1d96aa72–2f98–11e3-8906-3daa2bc de110_story.html.

4   *Cosmic inflation*: The development of the inflationary uni-verse theory was described by its originator, Alan Guth, in *The Inflationary Universe: The Quest for a New Theory of Cosmic Origins* (New York: Perseus, 1997).

5   *Inflatons*: Not much is certain about the hypothetical pro-posal of inflatons, which is probably why so few popular-level accounts exists. Your jumping-off point is the following: Paul Sutter, "How Did Inflation Happen—and Why Do We Care?," Space, October 26, 2018, https://www.space.com/42261-how -did-inflation-happen-anyway.html.

6   *Exotic phases of matter*: Beyond liquid, gas, and solid are plasma and many exotic phases that researchers are yet to come to grips

with. See Natalie Wolchover, "Physicists Aim to Classify All Possible Phases of Matter," *Quanta Magazine*, January 3, 2018, https://www.quantamagazine.org/physicists-aim-to-classify-all-possible-phases-of-matter-20180103/.

7 *Supercooled water*: If you don't want to do the experiment yourself, type the term into the YouTube search field and find hundreds of videos of instantly freezing water. We recommend this one: D. Muller [Veritasium], "Supercooled Water—Explained!," March 22, 2011, YouTube video, 3:35, https://youtu.be/ph8xusY3GTM.

8 *On the matter of dark matter*: A lucid explanation of dark matter comes from its tenuous connection to the extinction of the dinosaurs of all things! See Lisa Randall, *Dark Matter and the Dinosaurs: The Astounding Interconnectedness of the Universe* (New York: Harper Collins, 2015).

Why is there matter in the universe?

1 *Dirac and the antiparticle*: For more, try this award-winning story of his life and his prediction of antimatter: Graham Farmelo, *The Strangest Man: The Hidden Life of Paul Dirac, Quantum Genius* (London: Faber and Faber, 2009).

2 *Funnily named quarks*: Quarks were named by Murray Gell-Mann as an arbitrary whimsy that stuck, grabbing the word from James Joyce's *Finnegan's Wake*. Read more at Susan Kruglinski, "The Man Who Found Quarks and Made Sense

of the Universe," *Discover*, March 16, 2009, https://www.discovermagazine.com/the-sciences/the-man-who-found-quarks-and-made-sense-of-the-universe. There appears to be a transatlantic divide on the pronunciation of quark, with those in the United Kingdom rhyming it with "Mark," while in the United States, it rhymes with "Mork."

3   *Symmetry is central to much of fundamental physics*: An excellent discussion, including the importance of symmetry breaking, can be found in Martin Gardner, *The Ambidextrous Universe* (New York: Basic Books, 1964).

4   *More on Noether*: Emmy Noether is one of the truly unsung heroes of modern science and mathematics. Her insights now sit at the core of much of physics, and her story deserves to be known more broadly. See Dwight E. Neuenschwander, *Emmy Noether's Wonderful Theorem* (Baltimore: Johns Hopkins University Press, 2017).

5   *The Wu experiment*: Chien-Shiung Wu's success as an experimental physicist earned her many nicknames, which are repeated in bibliographies and even children's books. See Teresa Robeson, *Queen of Physics: How Wu Chien Shiung Helped Unlock the Secrets of the Atom* (New York: Sterling Children's Books, 2019).

Where did the elements come from?

1   *Atomic structures*: Rutherford's discovery of the structure of the atom is nicely described in Brian Cathcart, *The Fly in the Cathedral: How a Small Group of Cambridge Scientists Won the Race to Split the Atom* (London: Penguin, 2004).

2   *Problems in creating the elements*: You can play with nucleosynthesis in the early universe with AlterBBN, a program that computes the abundances of the elements predicted to have existed after the Big Bang (https://alterbbn.hepforge.org). With full documentation, you can understand just how cosmologists calculate the initial mix of elements in the universe.

3   *Pauli and his famous principle*: A succinct historical account of the exclusion principle appeared in the newsletter of the American Physical Society: "This Month in Physics History: January 1925: Wolfgang Pauli Announces the Exclusion Principle," *APS News* 16, no. 1 (January 2007), https://www.aps.org/publications/apsnews/200701/history.cfm.

4   *Braaaaaains!* There is a lot of serious academic research on zombies. The reason is that popular fascinations are more engaging than abstract models. See Andrew Trounson, "The Hard Science Behind Surviving a Zombie Attack," *Pursuit*, May 22, 2018. https://pursuit.unimelb.edu.au/articles/the-hard-science-behind-surviving-a-zombie-attack.

PART 2: THE QUANTUM OF COSMOS PRESENT

How did we unravel the chemistry of the heavens?

1   *The birth of astrophysics*: A detailed discussion of how we went from telescopes to cosmology in a single century can be found in Malcolm Longair, *The Cosmic Century: A History of Astrophysics and Cosmology* (Cambridge: Cambridge University Press, 2006).

2   *Original source*: Auguste Comte, *Cours de Philosophie Positive, Tome II* (Paris: Bachelier, 1835).

3   *Disappearing helium*: On the ever-depleting resource essential for science, see Michael Greshko, "We Discovered Helium 150 Years Ago. Are We Running Out?," *National Geographic*, August 17, 2018, https://www.nationalgeographic.com/science /2018/08/news-helium-mri-superconducting-markets-reserve -technology/.

4   *Original source*: "The Nobel Prize in Physics 1918," Nobel Media AB 2020, December 4, 2020, https://www.nobelprize .org/prizes/physics/1918/summary/.

5   *The birth of the quantum*: A detailed historical account of Planck's quantum hypothesis can be read in Clayton A. Gearhart, "Planck, the Quantum, and the Historians," *Physics in Perspective* 4 (May 2002): 170–215, https://doi.org/10.1007 /s00016-002-8363-7.

Where did the chemicals inside us come from?

1   *More playing with nuclear energy*: If you would like to play with stellar evolution, you should try our Modules for Experiments in Stellar Astrophysics (http://mesa.sourceforge.net).

2   *Forming the elements*: A popular discussion of the formation of elements can be found in Marcus Chown, *The Magic Furnace: The Search for the Origins of Atoms* (Oxford: Oxford University Press, 2011).

3   *Tunneling into the past*: Quantum tunneling is a ubiquitous concept used in all application areas of quantum physics. For a history, see Eugen Merzbacher, "The Early History of Quantum Tunneling," *Physics Today* 55, no. 8 (August 2002): 44, https://doi.org/10.1063/1.1510281.

4   *What happens inside the tunnel?*: It was only this year that physicists measured the time it takes a particle to tunnel, proving that it spends time while inside a barrier. See Ramón Ramos, David Spierings, Isabelle Racicot, and Aephraim M. Steinberg, "Measurement of the Time Spent by a Tunnelling Atom within the Barrier Region," *Nature* 583 (2020): 529–32, https://doi.org/10.1038/s41586-020-2490-7.

5   *Fred Hoyle had a huge impact on twentieth-century astronomy*: Read more about him and his life in his autobiography: Fred Hoyle, *Home Is Where the Wind Blows: Chapters from a Cosmologist's Life* (1994; repr., Mill Valley, CA: University Science Books, 2015).

Why do dying stars rip themselves apart?

1   *The characteristics of stars*: There are many ways to classify stars. One of the most popular, called the Harvard spectral classification system, was developed by Annie Jump Cannon and arranges stars by temperature. Our star, the Sun, is a class G star in this system, for example. Cannon classified a staggering 225,000 stars. See *Encyclopaedia Britannica*, s.v. "Annie Jump Cannon," accessed November 24, 2020, https://www.britannica.com/biography/Annie-Jump-Cannon.

2   *The story of the neutrino*: A popular account of the history of the neutrino can be found in Ray Jayawardhana, *The Neutrino Hunters: The Chase for the Ghost Particle and the Secrets of the Universe* (London: Oneworld Publications, 2014).

3   *Detecting neutrinos*: Science experiments only get bigger. The next generation neutrino detector, P-ONE, will sit on the sea floor of the Pacific Ocean. See Edwin Cartlidge, "Astronomers Plan Huge Neutrino Observatory in the Pacific Ocean," *Physics World*, September 18, 2020, https://physicsworld.com/a/astronomers-plan-huge-neutrino-observatory-in-the-pacific-ocean/.

4   *SN1987A*: We've been watching a supernova happen for over thirty years now, and it is beautiful. The name also makes it easy to search the web for. Not everything you find when searching for "supernova" will be scientific, but "SN1987A" doesn't have enough of a ring to it to be corrupted by the weirdness of

the internet. We recommend starting here: "Supernova 1987A Illuminates after 30 Years," Australian Astronomical Optics, February 23, 2017, https://www.aao.gov.au/news-media/media -releases/Supernova1987A-30.

Is the entire universe a quantum thing?

1   *Exorcising a demon*: Like Schrödinger's infamous cat, Maxwell's demon is also the subject of much popular culture. However, if you want a popular science account of the demon and its relation to everyday life, check out Hans Christian von Baeyer, *Maxwell's Demon: Why Warmth Disperses and Time Passes* (New York: Random House, 1999).

2   *More information on information*: Maxwell's demon, like an ever-increasing amount of physics, has been understood from the point of view of information theory. For an excellent account of what exactly information is, see James Gleick, *The Information: A History, a Theory, a Flood* (New York: Pantheon Books, 2011).

3   *Interpreting quantum interpreters*: In the over ninety years since the first official interpretation of quantum physics, the best description of what is going on is in a YouTube video: D. Walliman [DoS—Domain of Science], "The Interpretations of Quantum Mechanics," April 3, 2019, YouTube video, 17:11, https://youtu.be/mqofuYCz9gs.

4   *Why argue so much about a theory that works?*: For a deeper

discussion of quantum interpretations and other foundational issues in physics, see Lucien Hardy and Robert Spekkens, "Why Physics Needs Quantum Foundations," *Physics in Canada* 66, no. 2 (2010): 73–76, https://arxiv.org/abs/1003.5008.

5   *Many worlds, many books*: In popular culture, the many-worlds interpretation is the most famous, probably since it makes for good science fiction. A recent defense of this interpretation was given in lucid fashion by Sean M. Carroll in *Something Deeply Hidden: Quantum Worlds and the Emergence of Spacetime* (New York: Dutton, 2019).

PART 3: THE QUANTUM OF COSMOS FUTURE

Why don't all dead stars become black holes?

1   *The death of stars*: For a scientifically accurate video, including real stellar imagery, see Oli Usher, "Hubblecast 52: The Death of Stars," ESA/Hubble, January 17, 2012, video, 6:49, https://www.spacetelescope.org/videos/hubblecast52a/.

2   *Chandrasekhar unlimited*: Subrahmanyan "Chandra" Chandrasekhar won the 1983 Nobel Prize in Physics for his work on stellar evolution, including the 1.4 solar mass limit of white dwarf stars that bear his name. But he also published more than 10 textbooks and 380 academic journal articles while supervising the theses of more than 45 PhD students. How Chandra made it from mid-century India through

England to America is a fascinating tale told in Arthur I. Miller, *Empire of the Stars: Friendship, Obsession, and Betrayal in the Quest for Black Holes* (New York: Houghton Mifflin Harcourt, 2005).

Will matter last forever?

1   *Nobel Prize(s)*: Marie Skłodowska Curie is the only person to win multiple Nobel Prizes in different scientific categories (Physics in 1903 and Chemistry in 1911). There is no shortage of biographies, biopics, and dramatizations of her life. A unique YouTube video outlines the PhD thesis of M. Skłodowska Curie in cosplay, which we highly recommend: Toby Hendy [Tibees], "Marie Curie's PhD thesis ⚛," June 4, 2020, YouTube video, 14:09, https://youtu.be/-Vynhniw7SY.

Do black holes last forever?

1   *Paradoxical black holes*: Where the evaporation of black holes and conservation of information collide, physicists hope to find answers about quantum gravity. A recent overview with great infographics is here: George Musser, "The Most Famous Paradox in Physics Nears Its End," *Quanta Magazine*, October 29, 2020, https://www.quantamagazine.org/the-black-hole -information-paradox-comes-to-an-end-20201029.

2   *Black hole evaporation*: We know some of our physics colleagues will be rolling their eyes with the use of this analogy.

Hawking drew this very picture of virtual particle/antiparticle pairs at the horizon in his extremely successful book, *A Brief History of Time: From the Big Bang to Black Holes* (New York: Bantam, 1988). But this is a picture of some extremely complex mathematics, and explaining who sees a virtual particle as being virtual and who sees it as real is a topic too large for this book. For the sake of brevity—and sanity—let's stick with Hawking's analogy. He did okay with it.

Is the end of the universe really the end?

1   *The end of the universe in a little more than 280 characters*: Katie Mack, whom you should follow on Twitter at @AstroKatie, tours the possible final states of the universe in *The End of Everything: (Astrophysically Speaking)* (New York: Scribner, 2020).

2   *Video killed the book star*: If you would like to watch one of the authors speaking about the future of the universe, go here: Royal Institution, "The End of the Universe—with Geraint Lewis," October 3, 2018, YouTube video, 57:48, https://youtu .be/IF4UhElRUFg.

3   *Arrows of time*: There are more than a few proposals for where time gets it direction. An annotated bibliography is here: James B. Hartle, "Arrows of Time," The Quantum Universe, accessed November 25, 2020, http://web.physics.ucsb.edu/~quniverse /arrows.html.

4   *Darker and darker*: The descent of the universe, from glorious starlight today to the blackness of an unending tomorrow, is charted through *The Five Ages of the Universe: Inside the Physics of Eternity* by Fred Adams and Greg Laughlin (New York: Touchstone, 1999).

PART 4: THE FUTURE OF A QUANTUM COSMOS

What does a theory of everything look like?

1   *Theories, theories, theories*: For a rundown of the most popular candidate theories of everything, see Michael Marshall, "Knowing the Mind of God: Seven Theories of Everything," *New Scientist*, March 4, 2010, https://www.newscientist.com /article/dn18612-knowing-the-mind-of-god-seven-theories -of-everything/.

Index

Note: Numbers in *italics* indicate illustrations.

Q

About the Authors

CHRIS FERRIE is an associate professor at the University of Technology Sydney in Australia, where he researches and lectures on quantum physics, computation, and engineering. He is the author of over fifty children's books about science, including *Quantum Physics for Babies* and *There Was a Black Hole That Swallowed the Universe.* As the father of four curious children, he believes it is never too early to introduce kids to big ideas!

GERAINT F. LEWIS is a professor of astrophysics at the University of Sydney, where he hunts for the dark side of the cosmos, the mass and energy that dominate the universe. He teaches physics and cosmology, and he is the author of two popular-level books on the universe and how science works. He regularly speaks on the wonder of the universe to international audiences. His favorite fundamental force is the weak force!